The Ju-Jitsu of the Peahen

The Ju-Jitsu of the Peahen

Illustrated Essays on Death, Dying, Altruism and Natural Selection from an Unlikely Natural Historian

A.B. Williams, M.D.

Dedication

For my wife, Alaine.

Ju-Jitsu, Jujitsu, or Jujutsu

The guiding principle behind Ju is the idea of a weaker (gentler) force overcoming a stronger force through the application of technique, or Jitsu, rather than through strength and aggression.

From *Mastering Jujitsu* in <u>Human Kinetics</u> by Renzo Gracie and John Danaher

The views expressed herein are solely the views of the author and do not represent the views of his employers.

Table of Contents

Preface

Why do we die? A question pondered by any self-aware individual, surfacing once death is recognized as an eventuality. It has been asked throughout the millennia and answered, if unsatisfactorily, in myth, religion and recently in scientific epistemologies. As surgeons, the grim possibility of death, that ubiquitous sword of Damocles, hangs over the countless weighty decisions we are called upon to make. And for those weakened by illness who await the outcome of these decisions, the horsehair that holds aloft the heavy sword does not hold much faster than the tenuous tendrils belaying death. Why has nature seen fit to vex, sometimes torment, us with our certain mortality? The answers seem lacking in terms of natural selection; the truths held sacrosanct as handed down by Darwin (that traits that lead to more reproductive success should be passed on more often and be more prevalent amongst species) should, over the epochs, lead to an individual that is long-lived and quite possibly immortal. How do we jibe this with the fact that no species yet identified lives indefinitely? After all, the pool of life on Earth is capable of regenerating new tissue and creating new and vigorous entities. Why not rechannel this energy into extending our reproductive life-span and longevity in general?

The Ancient Greeks embodied longevity as a tempestuous sorority of three, known as The Fates. The three sisters spun, measured, and cut the thread of life. The first sister, Clotho, spun the golden thread of life. The second, Lachesis, measured the length and determined one's destiny. Atropos, or she who cannot be turned, cut the thread of life whence death ensued. This textile triumfemate was more powerful than the gods; even Zeus had to accept their ministrations. Modern evolutionary biology has a multitude of theories of why we die and why we age, from a gene-centered explanation to a theory based on multi-level fitness. Are we better off than the Greeks who "knew" why we die? As Lee Segall said, "A man with one watch knows what time it is; a man with two watches is never quite sure."

For the most part this work is opinion. And, at the risk of being crude, opinions are like gas; everyone expels theirs from time to time and no one thinks their own stinks. I offer no new data to interpret, just conjecture, hypothetical scenarios and a circumstantial case for the reason we die, the reason we are unselfish and tangentially why the peacock has such a refined caudal embellishment. There are other tidbits as well, absconded from my familiarity with the frailties of human physiology.

This book is written to be enjoyed by scientist and nonscientist alike. It is part scientific paper, part thought experiment and part short story anthology. As a physician, my credentials as an evolutionary scientist are too meager to be rigorously relevant. Hopefully my ministrations are neither too pedantic for the professional ethologist nor too esoteric for the casual evolutionary enthusiast. If you fall into the latter, enjoy the last half of the book; the former should dissect the first half. Please don't get caught up in remonstrating my short comings as a biologist; I am quite sure there are many. Instead focus on the larger message, that natural selection can leave us with seemingly antithetical processes of selection. Though thoroughly researched, the topics, I must admit, are a little bit out of my bailiwick; hopefully I haven't unwittingly stepped on any toes or hopelessly misconstrued previous works. This work flowed from a thought experiment of my own deliberations; I tried to avoid dogmatic furrowing and promote unconventional thought. Support has been difficult to obtain for an unconventional idea from an unlikely naturalist. That said, I relied heavily on the works of Richard Dawkins, Steven Jay Gould, E.O. Wilson, Amotz Zahavi and online compendiums for help when stuck in rut. Furthermore, the patients detailed in the examples are amalgams of many patients and are purely fictional. Any resemblance to actual persons is purely coincidental. The whole of this work may not add much to mainstream evolutionary thought but may stimulate others in the field to look at evolution in a different light. Perhaps it will poke holes in the fences between group selectionists and gene centrists. Evolutionary thought has been stymied by the entrenched positions of the groups involved; perhaps this book will render the arguments less vituperative. Perhaps it will only add to more dissension. Take from it what you will and let me know what you think.

www.peahenjujitsu.com

PART I

A Tale of Two Theories: The Gene vs. Something Larger

Introduction

From the Guinness Book of World Records, the modern record for human longevity that has been verified is 122 years and 164 days. The oldest mammal on record is thought to be a bowhead whale, aged from 225 to 250 years old. Impressively long lived, no doubt, but not immortal. Certain invertebrates seem to fare better than we terrestrial hot bloods in the sphere of longevity. One clam genus has been documented to live as long as 400 years. For some plants, longevity is downright Methuselistic. The oldest organism currently is a bristlecone pine thought to be over 4,800 years old.

Figure 1. Bristlecone pine forest in Inyo National Forest, California/USA. Photo courtesy of Wikipedia. Photographed by Siehe Unten.

Why do we die? All individuals of a species are mortal. We are capable of creating youthful tissue (at least the female of our species is capable with a little help) so why then have we not as a species evolved to stay young and vigorous, to replace worn out proteins and tissues and live for eons? Our liver, for example, if not overly abused, could possibly be immortal if the rest of us didn't fall apart around it. Some reptiles have an unending supply of teeth and as a class are capable of regenerating new appendages. Wouldn't the processes of evolution favor the long-lived who had a chance to reproduce more often?

Well, for one, perhaps we are simply thermodynamically challenged. Thomas Kirkwood explained in 1977 that from the perspective of natural selection, once we have passed on our genetic material, our bodies are disposable. As warm-blooded entities, we are constantly exposed to entropic attack at the molecular level. In other words, just as my teenager's room doesn't stay neat and ordered for long, so goeth our neatly ordered, structured physiology. Our main source of fuel, oxygen, itself is a cellular wrecking ball, once properly metabolized. Life at 98.6° F is akin to a slow simmer bound to cook us all eventually. Fats, which we Americans have in abundance, are treasure troves of electrons and ionizing energy that, once unshackled, are like loose cannons on a man-o-war. Throw in ultraviolet light, incessant gravity, smog, pollutants, environmental hazards, fried chicken and the odd brother-in-law and it's a wonder we don't age faster. We simply break down in ways that cannot be repaired. Cold blooded animals presumably fare better from the inward metabolic attack at a cooler temperature and can live longer, but eventually they too succumb to the slings and arrows of outrageous enthalpy.

Compounding the search for the molecular Fountain of Youth, the betterment of individuals afforded by natural selection seemingly applies only to the young and fecund; natural selection is a potent sculptor of adaptation but only capable of shaping marble that is young in age. Natural selection shapes a species form and function up to the age of reproduction; what longevity that occurs past fecundity is gravy on your mashed potatoes. Tattered and tired by life's hurdles to pass the baton in our species' genetic relay race, it seems we are loosed from the ceaseless honing of Darwin's grindstone once the last of our offspring are independent. Or is this entirely correct?

Another facet of the answer of why we die is more subtle and perhaps counterintuitive at first. However, we are evolutionarily more successful if we, as individuals, don't live forever. Longevity brings wisdom, experience and more opportunities for the individual to procreate, but at a high cost. Longevity, in a resource constrained environment, means less food for the species as a whole. Environmental resources are not constant. It is a boom and bust cycle of punctuated abundance followed by, at times, prolonged scarcity. The species or group that over expands during good times jeopardizes the most important members of a species during times of scarcity. As a group, it may be prudent to keep numbers at or below the carrying capacity of the environment at its lowest point. Longevity puts the species as a whole at risk.

But there is a problem with this notion. The conventional wisdom surrounding evolutionary theory implies that natural selection does not work efficiently at the level of the group. It is perhaps one of the great misconceptions about evolution by natural selection, that we have evolved characteristics or traits that benefit the greater good. This may be a rung in the ladder of our moral evolution, to be compassionate, selfless and group centered; too often, this is confused with the course we feel natural selection should take.

The metaphor of natural selection and athletic competition is irresistible to most. Thus we should evolve towards cooperation, teamsmanship, and subvert individual achievement for the good of the team. Yet natural selection, red in tooth and claw (any book on natural selection is obliged to include Tennyson's catchphrase; there I've written it), is driven only by increased fitness of offspring, not any moral imperative. Groups with a certain moral trait competing against other groups without that trait, is too slow a process to be effective and is outmaneuvered by its constituent individual-based selection. In other words, individuals (or perhaps even individual genes) drive natural selection. Groups are too large, too heterogeneous, and too ponderous to be the arbiter of efficient natural selection. We may make alliances, but if these do not favor individual and offspring survival, the tendency toward making such alliances will be winnowed. When the goal is survival of our genetic material, we are more akin to cyclists than footballers.

But if individuals are the tip of the spear evolutionarily speaking, all things being equal, we should live a long time. Natural selection should favor longevity, even immortality; yet it does not in any species.[1] Alas, the devil is in the details; all things are never equal. There is a price to be paid by the gene, who longs to be widespread, if its carrier (the individual) lives forever.

Individuals of a species have discovered in Darwin's vast laboratory of trial and error, that individual longevity, in a general sense, is bad for gene propagation. In the context of natural selection, the gene that disperses too widely puts itself at risk. Because genes are ensconced in protoplasmic quanta dependent upon external energy resources, survival and dispersion of a gene don't solely depend on a more-is-better strategy. The gene must concern itself with the health of its carrier. If replication of a gene under conditions of resource scarcity means decreased fitness for its carriers, then selection mechanisms should arise to avoid gene replication under those conditions. By not replicating its carriers all the time, the gene is more fit. Thus there should be a magic gene number[2]- a Goldilocksian integer- not too many that too few survive and not too few that not enough survive, that can fluctuate depending on environmental conditions. While there are mechanisms to make sure that the number of "births" of gene carriers is regulated, we ought to expect that there would be selection toward regulating the number of "deaths" of gene carriers as well. After all, genes don't carry crystal balls. Individuals whose genes could affect not only the rate of births but also have some impact on rates of death would come closer to attaining that magic gene number.

I am a physician, a surgeon, more suited to the operating room than the fossil record room. Everyday though, I am confronted with our own species mortality. People die. Most of the time I prolong the inexorable journey the fates have prescribed to those under my care. But sometimes I shorten it. Hopefully not with any egregious act of negligence or carelessness, but through well meaning acts intended to help, but result in harm. The medical term is iatrogenesis. The road to hell is paved with good intentions. The intravenous line that I put in to administer life-sustaining nutrition becomes infected and the patient spirals down into kidney failure, pneumonia, and death. The operation I perform on the colon for cancer results in a serious skin infection that despite recognition and treatment develops resistance to antibiotics and the patient succumbs.

While struggling with these daily dilemmas, it struck me that we all die. While not a thunderous revelation, or any Archimedial eureka moment, it nonetheless is a ponderous, substantive reality that all men and women eventually die. All too often it is our own cells, tissues or organs that do us in. While studying human disease in medical school, I was struck by the fact that the body's compensatory efforts to fight disease often hasten death. When the heart starts to fail, for example, our small blood vessels constrict, our blood pressure rises, our kidneys hold on to more salt and fluid, our tissues swell and our lungs are filled with fluid. All of these things make it harder on the heart which is already failing. It's as if the body is saying "Uncle" at the first signs of adversity.

Certainly I am not the first to ponder these contradictions. A whole branch of thought dealing with senescence has developed to explain why we age and why each species has a unique, seemingly proscribed life history. Giants in this field are scientists like Peter Medawar, who recognized that effects of natural selection are magnified in the young and potentially fecund or George Williams who recognized that a gene's duplicitous effects are more important earlier in life than later on. Both argue that the effects of aging are detrimental to the individual and his potential genetic heritage. Medawar explains that late acting genes leading to death persist because of genetic drift. Because most in a species have died early due to predation or disease, there is no significant selection pressure for long lived genes. There is no pressure for genes to promote longevity if most are dead from extrinsic causes early in life. Williams explains that malignant genes later in life persist because they are more beneficial early in life. A gene can double cross its carrier, if its early effects are more beneficial (in a gene dispersion sense) than its late effects. A gene for efficient blood clotting, for example, would benefit the young and injured, but might cause the death of an older individual with hardened arteries. The sequence of the failing heart described above is in place to help the young who may have reversible low blood pressure from hemorrhage or dehydration. It is only on the aged that the duplicity of the system becomes apparent. I will try and argue that the late effects of aging and dying which are perceived as detrimental to an individual's genetic heritage, are actually beneficial; Medawar might argue that the genes responsible for aging might eventually be selected out, should individuals start living longer.

Though it could be delayed, I would argue that aging would be selected **for** because it is, paradoxically, vital for individual gene survival.

There are a host of other conundrums of human physiology that on the surface are extremely perplexing for amateur Darwinists like myself. Why do we have a gallbladder that can kill us in three or four different ways? Why does the appendix persist evolutionarily when its sole function seems to be to kill us? Why do our own cells rebel against us to cause incurable cancer or autoimmune disease? Why is self-destructive behavior still so prevalent? How has unselfish, self-detrimental behavior evolved? How has homosexuality, which, without procreation, is death of one's genetic material, persisted and flourished (pardon the pun)?

These phenomena seem to fly in the face of natural selection as proposed by Charles Darwin. We will look at more at this later, but from a crude level of understanding of natural selection or "survival of the fittest", as Herbert Spencer once coined, these self destructing mechanisms should have been weeded out as "unfit" characteristics, that should have been eliminated in the vastness of time of evolution's great laboratory. I will be the first to admit that I am an amateur evolutionary biologist; however, I do have a perspective on dying that, well, despite the overused colloquialism (and poor punmanship), is outside the box. Let's look at this more closely.

Chapter 1

GENE SELECTION vs. GROUP SELECTION

Evolution as proposed by Charles Darwin in 1859, essentially states that those individuals who are most likely to survive to the age of reproductive ability are most likely to pass on their superior survival traits. It relies on the random creation of new traits, or mutations, which in and of themselves may or may not be beneficial to the organism. If by chance, it is beneficial and aids survival, (say a longer neck in a giraffe-like ancestor) then that trait is passed on more frequently. If it is not beneficial, the individual is less likely to survive to maturity and less likely to pass on the trait. The trait is "selected" out, much like dog breeders select out certain traits by only breeding dogs with favorable traits (artificial selection).

This seemingly puts the individual of the species as the focus or tip of the spear of natural selection. In other words, to survive and for natural selection to occur, individuals are thought to compete against one another to both find resources and not become one. Indeed, this is what the word "selection" implies, a competition in which individuals are selected as winners. We are in competition with our fellow species mates to find food and not become food. The strongest, most cunning, most attractive, live and reproduce the most.

This idea, in all its permutations, invites confusion when attempting to explain some animals' behaviors or traits that would not appear to be beneficial for survival. It has trouble extricating why some individuals of many species exhibit altruistic or unselfish behavior. Birds and monkeys alert others to predators and call attention to themselves by calling out. They lessen their probability of survival by doing so, so shouldn't these unbeneficial traits be weeded out? Natural selection should lead to individuals that are completely selfish, who keep quiet and let others get eaten, or in other words, behave with their own survival as their tantamount concern. Why bother sharing a kill, why bother raising young, why bother feeding them while the parent goes hungry? Why even bother looking for a mate? Why not reproduce by cloning ourselves, instead of passing along only half of our hard fought, hard won, genetic material, which may or not be expressed in our offspring? Lastly, why not live forever, or at least minimize the deadly genetic landmines found throughout most species' genomes?

Richard Dawkins has examined these questions in his now famous book *The Selfish Gene.* He explains that altruistic behavior ultimately is a selfish endeavor, in that by acting altruistically it more often than not favors the individual through reciprocity and sneakiness. The "You scratch my back and I'll really, really scratch yours later" theory favors the individual who can game the relationship to increase his individual fitness. If one plays the game properly, it behooves an individual to act altruistically because ultimately, the energy benefits in the long term outweigh the energy inputs in the short term. Furthermore, acts that on the surface appear altruistic, actually improve the survival of the individual once all the layers of the behavior are peeled back. For instance, as Professor Dawkins (with a nod to Robert Trivers) explains, alert calls by birds can send the group into hiding before they are spotted by the predator. This allows the predator to pass on by, which improves everyone's odds of survival including the one who sounded the alert.

Furthermore, for Dawkins, it is the gene that is driving behavior, not the individual. The individual is simply the vehicle of the gene. As long as the trait improves the survival and propagation of the gene, it will be selected for, no matter how antithetical it may appear to our poor powers of calculation.

The strength of this theory lies in its primality. The prime mover in gene selection theory is DNA itself, which is the engine, the *sine qua non*, of natural selection. In the beginning, in the primeval soup of a proto-biotic earth, it was the stringing together of deoxy- or ribonucleic acids which could then not only replicate themselves, but also assemble other amino acids into proteins. This allowed natural selection to occur at the molecular level and for incipient life to take hold.

Its weakness is also its primality, in that greater than 99% of life on earth, the genetic material has become dependent upon the proteins they code for, the carrier in which they ride and other genes within the individual's genome. Just as the architect's design is constrained by the costs of supplies, the building material available, and the availability of his workforce, so the gene is confined within the constraints of the environment around itself. The coach of a football team might have a grand scheme, but because he is confined to the sidelines, he is wholly reliant upon his players and their talents to make him successful. If the coach pushes his players too hard they will be exhausted and not able to compete. Similarly, the gene cannot push too hard on its carriers. The gene is wedded and indebted to the individual carrying it; a gene cannot make too many copies of itself with little or no regard for its carrier.

In trying to explain why we die, it would be easy to invoke a group or multi-level selectionist's idea that we as members of a population or group, who die at a certain time, can outcompete other groups without such a gene for death. By staying lean as a group and focusing resources on the young and potentially fecund, it is easy to imagine this group outcompeting others more concerned with individual survival. However, that would be too easy and this work would be all but over. The problem with group selection is it simply takes too long to occur; its premise is false that groups are predominantly homogenous, and any trend toward communal behavior would be undone by individuals in the group who would benefit by selfish motives. There may be some aspects of group selection which may occur, but it is simply not a prime mover in the creation of new species or evolution in its intricacies.

If we look at this theoretical knot in a slightly different way, the back and forth debate of gene vs. individual vs. group selection could be made less diametric. The theory of group selection has arisen to explain traits that seem maladaptive for the individual, like altruism. Genes that maximize the survival of the majority of individuals in a brood, group or even a population at the expense of other individuals can be selected for if the overall effect of the gene is positive. If we can find an explanation for why a gene could seem maladaptive for some, but in actuality, is adaptive for individuals in the majority in the group, then there would be no need to invoke group selection at all as an explanation for the appearance and proliferation of that gene.

For instance, the gene for sickle-cell disease arose in a population where malaria was endemic. The malaria parasite replicates inside a human's red blood cells. Those with only one copy of the gene for sickle-cell disease are somewhat protected from the ravages of malaria, but otherwise are not seriously harmed by being carriers of one copy of the gene. Full blown (those with two copies of the gene) sickle-cell disease causes the red blood cells to be misshapen. Instead of looking like a small jelly donut pinched inward at the center (a biconcave disc) which can wiggle through the smallest of capillaries, the red blood cells instead are sickled, like a farmer's old fashioned harvesting implement, and stiff. They clog up the small capillaries throughout the body, leading to pain, poor circulation and ultimately death. Individuals with only one copy of the gene benefit by having the sickling gene in the gene pool, even if individuals with two copies of the gene (full sicklers) do not. On the surface, it might appear to be group selection; groups where some members have a maladaptive trait like sickle-cell disease out compete those groups without it. Of course it is not group selection; the gene flourishes because **individuals** with one copy of the gene are more fit than individuals without the gene in areas where

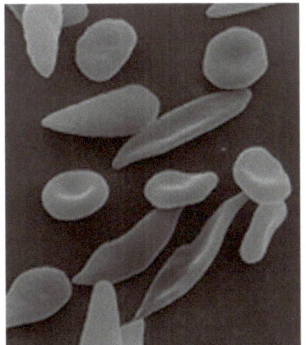

malaria is endemic. The partial sicklers are able to out compete individuals which lack the gene, or more accurately, the allele for sickle-cell, even if the full sicklers suffer. The overall effect of the gene is beneficial. If the gene for altruism were like the gene for sickle-cell disease, it could spread in a similar fashion. If one further peels back the onion, it appears that, within the framework of Mendelian genetic inheritance patterns, the gene or allele for sickle cell disease is exploiting itself. We shall examine this in depth later.

Figure 2 Normal red blood cells appear like a small jelly doughnut. Sickled red blood cells are elongated and misshapen. The photo is attributed to Drs. Noguchi, Rodgers, and Schechter of NIDDK.

Chapter 2

WIDESPREAD GENES

The premise of natural selection is that traits that provide an advantage to survival of themselves and their offspring will be favored over others in a population. Because traits are driven by genes, it stands that genes that provide an advantage will be favored over others in a population. Such a simple concept can have amazingly complex iterations. While it is easy to see why faster muscles might be an advantage, it is much less clear why other adaptations might be advantageous, like a peacock's tail or a gene for depression. This is because in trying to interpret the effects of natural selection we cannot see the entire movie. We only see a small snapshot of the effect of a trait.

For example, suppose you were an alien xenobiologist trying to interpret human evolution and you landed on earth on the outskirts of Hiroshima, Japan in August, 1945. You would witness the death of nearly 100,000 people who seemingly were simply going about their daily lives. You would see the destruction of an entire city remote from any battlefield and the annihilation of a seemingly peaceful town. You might quickly get back in your spaceship and assume this was unequivocally a malignant act.

On the surface, the atomic bomb was a malevolent implement; however it ended up shortening World War II and saved an estimated millions of Japanese and Allied lives. This we can interpret from the benefit of a detailed knowledge of the history of the war, the Japanese people and their diehard fighting élan, and decades of historical input. Without such knowledge, we would be flabbergasted that such a deadly event could have evolved in the history of mankind.

Interpreting the beneficence of evolved traits via natural selection is subject to some of the same pitfalls. Why a trait is advantageous can be very difficult to tease out. This is partly because our powers of reasoning are filtered through our own biases of good and evil. We assume it's beneficial for an individual to live a long life, but salmon, mayflies or other semelparous organisms would argue it's more adaptive to die relatively young. Mostly though, we just can't tell what's going on in a movie by looking at the poster or even seeing the trailer. We can't always reliably tell what a species' evolutionary strategy is by looking at an individual trait, given only a short amount of time to view it and given our poor power to confabulate a "just so" genetic tale.

Compounding this situation are the cycles of dearth, not only from week to week, say when trees finish fruiting, but from year to year, when prolonged droughts occur, that exposed our ancestors to resource uncertainty that is hard to imagine today. Utterly repugnant events today like fratricide, an offspring's early death, or indifferent parenting might have been adaptations that improved survival for our ancestral hominids. Other adaptations that seem to fly in the face of natural selection to today's interpreters may be understandable in the context of this crucible of environmental pressure that once molded our species' form. Even in historic times, acts of human sacrifice, capital punishment for meager offenses, and mass murder were and are all too common. We have to be open to the idea that death is not always a bad thing in the context of natural selection and gene propagation. Moral and religious astigmatisms we have evolved to cope with the stresses of our hypersocial existence in an era of unheard of resource abundance further impedes the rational interpretation of adaptations that would only make sense were we privy to reams and reams of population census data.

Nonetheless, it is fun to try.

One aspect of natural selection that is misleading is the concept that a gene seeks to be widely spread through its environment. The idea that a particularly adaptive gene will diffuse like a gas widely and relatively quickly belies what I think actually happens. It's the concept that if one animal with a certain trait has more offspring survive and pass on that trait than another, then that trait will be more widespread. Up to a point, this is true. However, the gene is wedded to the individual, its carrier. The gene's survival and propagation is dependent upon individual survival. If the number of any gene expands so much that it risks the health of the individuals that carry it, through too much competition for food or shelter, then that risks extinction of that gene (or at the very least, out competition by other less fruitful genes). I would argue that an "adaptive gene" seeks to be efficiently spread throughout its environment. Here's what I mean by that.

If natural selection were simply a numbers game, organisms would be reproducing all the time. Obviously this is not sustainable. There isn't enough energy in ecosystems to sustain such growth. So, at equilibrium, organisms would be expected to curtail their expansion by reducing the number of births. And indeed this is the case. Most animals have a short estrous periods, and depending upon the parenting strategy and predation effects, a clutch size that aims toward maximizing offspring survival and at the same time minimizing environmental pressure.

However, only regulating birth numbers is not very efficient. DNA may form crystals, but it has yet to form crystal balls. It might be practical to overshoot the "magic number" of births and let predation or starvation sort out the magic number, but this would be inefficient and wasteful. Inefficient, for example, because in trying to keep four offspring alive, the risk is, that during a drought or food scarcity, the environment which could normally sustain four, now can only sustain one. In this situation, all could potentially die. Furthermore, if the environment were unusually bountiful one year, it would penalize the individual that limits brood size. If there were a mechanism to get rid of others with competing genes, which turned on in times of food scarcity, (or better yet, even before food got scarce) that would be a more efficient mechanism than only regulating birth numbers, and would be selected for.

Indeed we have recognized some forms of these adaptations in nature. Infant fratricide is well documented among some bird species. The first to hatch will either get the lion's share of the food leaving the others to starve or will out and out peck the younger siblings to death. Infanticide in rats or hogs by the mother or in lions by a newly dominant male is also seen. But offspring do not only compete against their own siblings for survival, they also compete against every other member of their species struggling for the same resources. Cousins, neighbors, mom, dad, grandpa and grandma all vie for resources that the offspring need. Because offspring are relatively helpless but exceedingly important, natural selection has imbued the genetic make-up of populations with subtle but efficient mechanisms for tilting the playing field towards enhancing their survival.

It is easy to see natural selection at work against predation. Antelope become faster or jinkier, in the words of Professor Dawkins, in response to the depredation of their pursuers. Moths become camouflaged depending upon the nature of their surroundings. However, likely, the biggest threat to the antelope is not the cheetah or lion, but other antelope competing for the same resources. Animals spend more time and energy foraging than running from predators. While it is true that predation does result in the majority of causes of mortality in some species, it is lack of food or sheltering habitat (as a result of overcrowding) that makes prey animals vulnerable to predation in the first place.

It is easy to see antelopes developing adaptations to combat predation- nature red in tooth and claw- (there it is again!) but it is harder to appreciate organisms developing cidal adaptations to combat members of their own species. Sibling infanticide is well documented in birds. Though less recognized, it is no less effective in reducing population stress if a" killing gene" acts on members of the species who are least likely to successfully reproduce or who have already reproduced. Relying on predation, disease or starvation risks the well being of the newest offspring.

Again keep in mind the idea of a gene's "magic number". This seems like another description of group selection in disguise, where groups who keep their numbers low are more fit than groups that don't. But it is decidedly not group selection. Through the crucible of natural (or gene or individual) selection, genes that are potentially cidal have arisen in individuals to maximize the exploitation of the environment without outstripping their ability to survive in it. This is especially important for species with little predation and especially important in environments where species have established a homeostatic balance between population number and resource availability.

One way to combat population pressure is out and out homicide; Konrad Lorenz famously documented this idea in his work *On Aggression*. However not all species are equipped for murder, and for social animals, its destabilizing influence is decidedly bad for group cohesion if it occurs within an animal's group; it is risky as well if aggression is directed towards animals of the same species outside the group when the outcome of the aggression is not in favor of the initiator. To combat population pressure, animals need a subtle mechanism that slowly but inexorably tilts the scales towards those with the most potential to pass on their genetic inheritances.

One such subtle mechanism is genetically programmed death. A gene or group of genes that brings about its own death, after its carrier has passed on its genetic material, is able to proliferate and survive, ironically because it kills itself. The fully independent offspring of the individuals with that gene(s) are better off without the parent or grandparent around. When and if the genetically programmed death occurs depends upon the species number, the carrying capacity of the environment, the effects of predation (and other extrinsic causes of mortality) on the species, and the parenting strategy of the species in question. In other words, salmon blatantly have the adaptation of death after reproduction but turtles don't, despite similar parenting strategies, is due to the fact that salmon numbers are far greater than turtles, making overpopulation more likely for salmon (especially before the days of commercial fisheries). For the salmon roe, the gene for programmed death of their parents (which they have inherited) is a trait the helpless roe use to exploit their own parents; that it will be used against them later in life is simply a consequence of the survival strategy they have successfully evolved. It's as if the salmon roe decided we would be better off without our parents lingering on, stripping the environment of what little resources are available. Since I don't have a gene that will off my parents, I will do the next best thing; I will off myself after my offspring are independent. The suicidal gene allows the roe to exploit their parents for their benefit.

But actually reducing gene numbers is tricky business for a successful gene. Well, it's not tricky for the gene, but it is difficult for us to interpret the effects of a gene that is potentially suicidal for its carrier, in the fabric of natural selection. Natural selection should outlaw genes that reduce individual fitness and in so doing, reduce the gene number.

The notion of exploitation, turned on its head, is sacrifice. From the perspective of the child, the death of the parent is exploitive; but from the point of view of the parent, the notion is sacrificial. The parents sacrifice their genes for the good of the genes in the offspring. This arrangement makes sense evolutionarily because more offspring survive because the parents die after spawning. It's an evolutionary two step for the gene: two steps forward (for the gene in the offsprings' generation) and one step back (for the parents'); the net gain in surviving offspring (and thus gene number) is more when compared to the scenario when the parents didn't die after spawning. Or at least it is not less.

Tangentially, this notion of dying for the good of one's offspring sounds a lot like altruism or kin selection. I think every parent would assent that we are all dying, slowly but inexorably, for the well being of our children. Whether it be the dangerous process of birth, the sharing of fat stores with offspring, risking life and limb in bringing home the bacon, these are all ways in which we decrease our own fitness for the good of our offspring.

This idea that we sacrifice for our offspring can be explained by kin selection. As W.D. Hamilton explained, because we share half of our genes with our offspring we are likely to be invested in their survival. There is even an equation to describe how invested we are in the survival of a relative depending upon the number of genes we share. However, kin selection has a hard time explaining why animals do things for strangers or for those not related to them. Why vampire bats will sometimes feed a hungry unrelated bat from the blood meal in their stomach. Why some species will adopt orphaned strangers. Why cuckoos are able to exploit other birds to raise their chicks. Why birds will call out when predators near an unwary bird, and why insect societies have developed sterile offspring who work for the good of the group.[3]

Furthermore, for kin selection to be effective the degree of relatedness of an individual needs to be ascertained. It may be difficult to tell who is related to you. Close relations may be identifiable, but it might be tough to recognize kin especially over time and in animals that migrate a long distance. It has been shown that kin selection works without having to recognize kinship so long as the behavior occurs in proximity to the animal in question. This assumes that relatives tend to live close to one another which for migratory or solitary species, seems like a contestable assumption.

It might make sense if these altruistic tendencies arose as an adaptation to subtly reduce population pressure in general and allow the realization of a gene's magic number. In other words, instead of looking for an advantage for the altruistic behavior, in reciprocal altruistic events later in life, in the complexities of game theory, or in kinship survivability, accept that the behavior is a disadvantage to the altruistic individual, but is an advantage for the individuals in the species who have selected for that gene. Most of the time, this exploitive/sacrificial behavior is geared towards one's offspring because of their importance as potential genetic torch bearers. But sometimes mature individuals can exploit members of their own species not related to them. Or, more fundamentally, perhaps there are ways in which a gene can exploit itself. To better understand this notion we have to look at the peacock's tail in a whole new light.

Chapter 3

THE PEACOCK'S TALE

For certain bird species differences in feather pattern, coloration and size between sexes are marked, unwieldy and potentially antithetical to evolution by natural selection. Charles Darwin recognized this dilemma. He noted two patterns of sexual dimorphism and postulated his own mechanism for their appearance. **Intra**sexual selection points to characteristics that develop to help the possessor fend off members of his own sex and win more mating opportunities. Antlers in elk species, horns in sheep and body variation in elephant seals for example, allow the male to fend off rivals and ensure his paternity. **Inter**sexual selection describes the development of characteristics that the opposite sex finds attractive and enhances the sexual fitness of the bearer of that trait (plumage, coloration, nest building). The latter traits, however, can be detrimental to the overall fitness of the individuals who possess them. The expression of these genes at sexual maturity leads to a male phenotype that is distinctly different than the female and in most cases hazardous to the longevity of the possessor.

Some biologists assert that intersexual selection is a result of the choice of the female, driven by the increased fitness that must be inherent in a male with such an impediment. In other words, having such a large tail makes the peacock egregiously and inimitably fit. This handicap principle, postulated first by Amotz Zahavi, states that the bearer of such an impediment will have other superior survival genes, like larger, faster muscles or quicker wits, to pass on. Hens that choose peacocks with incrementally larger tails will have male (and female) offspring who are also super fit and can out compete lesser endowed males. The peacock flaunts his undeniable vigor through his tail.

Another explanation, the "sexy son" hypothesis, first proposed by P. J. Weatherhead and R. J. Robertson in 1979, allows that it is likely advantageous to have offspring who are attractive to others within the species. It behooves the female to have sons with longer and longer tail feathers because he will have the most mating opportunities. Once the hens decided that a larger tail was attractive, a positive feedback loop was established that has resulted in the present day encumbrance. What's missing from this explanation is why a longer tail is so darn attractive for peahens. Wouldn't it be better if fashion jibed with function? Efficient, short tails that didn't result in the cock's early demise should be all the rage.

A third hypothesis is that the color and appearance of the peacock's train advertises his own health, how capable he is at resisting parasites. Just as a nice suit reflects the potential of a man on a first date, so reflects the vigor and constitution of the peacock through his bright and natty tail. However, experimental data hasn't always equated tail showiness with a bird's parasite load.

Another theory suggests that the origins of sexual selection are due to inherent tendencies of the peahen brain to select certain traits. Peahens might be stimulated by the color red if it reminds them of succulent berries or fruits. Or perhaps they are simply mesmerized by the intricate movements of the cock's tail shaking because it stimulates the neurons that already have been hardwired for survival. The bigger the tail, the more stimulation it provokes.

Paradoxically, the development of these dimorphic traits (under the influence of sex hormones[4]) will imbue the male offspring with more aggression, less attentiveness to his own chicks, a tendency towards expending energy involved in copulation (thus less energy for self-survival), louder vocalizations (attracting more predators), a weaker immune system, and a shorter lifespan.[5] The trade off for the male is to live shorter but have more opportunities for reproduction. Most explanations of the development of sexual differences have proponents and detractors because all of the hypotheses have weaknesses. None can adequately explain the benefit of evolving a trait that potentially curtails fecundity.

The attractiveness of a male peacock seems to correlate with how likely he is to die after copulation. The large encumbrance, the weakened immune system, and the louder vocalizations make him prone to early death. Perhaps, even Professor Zahavi would admit that the peacock, despite his undeniably manly compensations for the gaudy tail, still has a handicap that might curtail future breeding opportunities. This is, on the surface, is clearly mal-adaptive for the male[6]. But it might be adaptive for his mate and offspring.

Robert Trivers has shown that the more dimorphic the male of any species the less likely he is to participate in child rearing.[7] The peacock contributes nothing but gametes to his partner and fledgling brood. If the male is not likely to participate in any significant way to child rearing, then the female would be better off with the male dying shortly after fertilization so that she and her chicks will be better fed. For not only do she and her brood compete for resources with the cock himself, but also with every other potential offspring that he should happen to sire. The peahen is simply a little more subtle than the female praying mantis, which kills her mate immediately after copulation.

Is this then the stimulus for sexual selection? Peahens find a gaudy tail sexy not because it is subject to the whims of peahen fashion sense, nor because it advertises a healthy specimen underneath but because natural selection (not sexual selection) fostered the survival of the progeny of hens who chose fit peacocks who are nonetheless prone to dying and lessening the struggle of her offspring. I think most would admit that there is a number of surviving offspring for the cock, which should he surpass, would be detrimental for all past, present and future offspring. The peahen and natural selection are helping him find that number.

The peacock has been hoist by his own petard[8]; he goes along with the arrangement to the benefit of his offspring and because he has no other choice. Ironically, the cock's inclusive fitness improves by dying; his current offspring are more likely to survive by penalizing his own future fecundity. The bird at hand is worth more than the two birds in the future metaphysical bush.

For species that require the male to participate in child rearing there is less likely to be the dimorphism present in peafowl. If the female bird needs the male bird around for child rearing, natural selection is not going to favor animals with gaudy detrimental displays driven by testosterone. Raptors and other birds of prey with co-parenting strategies have little or no overt dimorphism (or have dimorphism favoring larger females who presumably will be doing the lion's share of the protecting but need the male about to share in chick rearing). In fact, the degree of sexual dimorphism in birds correlates well with the degree of paternal investment.[9]

This hypothesis that natural selection not sexual selection is driving sexual dimorphism in birds heuristically satisfies Occam's razor. Sexual selection relies on the likes or dislikes of the peahen to choose a mate. One is relying on the collective whims of potentially capricious hens to drive the evolution of the dimorphic species. Why should a slightly longer tail in ancestral fowl be selected for? Beauty may be in the eye of the beholder yet differences from the norm are likely to be unattractive. Perhaps longer tail feathers translate to a healthier bird, but incremental changes in feather length would be a poor metric for overall health. More likely, the sexual dimorphism present in certain species of birds is due to natural selection. Females with philandering, "deadbeat" males employ a strategy of selecting fit mates whose impediment hastens their demise after fertilization.[10] Incremental increases in tail length would be a good metric for increasing the probability of death. The bigger the tail, the more gaudy the coloration, the louder, more vibrant the song, or the more energy expended into non-survival activity, the more likely he is to die from predation, and thus the more attractive he is to the female. Females can exploit the male who "chooses" not to help with child rearing, by selecting traits in the male that are likely to benefit her and her offspring right now (as in birds with ritual nest making) or in the future by increasing the probability of death of the "deadbeat" male. Females who use this strategy will have more offspring survival, especially in a resource constrained environment. Conversely, species like the swan, penguins, or birds of prey, where paternal investment in offspring is high, will have the lowest degree of sexual dimorphism.

The degree of dimorphism that has evolved in certain species may depend upon mating strategy, paternal input on chick rearing, predation, environmental resources, and other intrinsic causes of death for the species in question. Coloration dimorphism, present in Mallards or other duck species, may be enough to slightly increase the mortality of males compared to females; any trend towards a more significant impediment in the males of these species might lead to a decrement in population numbers due to the species vulnerability in the aquatic environment. In other words, because the drakes are essentially serially monogamous but do not contribute significantly to offspring survival after mating, the Mallard hens do not select for an exuberant impediment because they have evolved a stable ratio of the sexes that requires access to an equal number of drakes. Were they to select for an audacious tail in the drake, there likely would be females who would not have a partner once mating season was completed. The increase in mortality of being a single parent mother might foster the evolution of a coloration pattern of the drake that made him as vulnerable to mortality from predation as a mother Mallard with a large brood of ducklings (who both remain camouflaged) to lay, hatch and protect. In this way, the ratio of sexes remains optimal for procreation and offspring survival.

Species tend to expand their numbers to the bounds of what the environment can support and predation tends to follow suit. In other words, at equilibrium, resources are scarce and predation isn't. If predation were such that the male added little to the protection of the chicks, and the chicks more or less survived by chance or by seclusion, then the males might find that they do better by not being a stay at home dad. They are better off making more offspring with as many females as possible. We should expect the females to really require an impediment for these males and be super selective as to which male should be chosen. For a hen to improve her brood's chances in this arena where food is scarce and predators abundant, she has to ensure that her brood has minimal competition and is maximally fit. She does so by minimizing the number of males available for copulation and maximizing the fitness of the males that are available. For the peahen, who requires the outlandish impediment in the cock, she ensures her brood will inherit powerful muscles and quick wits for escaping predation and either not inherit the impediment in half her brood (the females) or not inherit the impediment until they are sexually mature.

Furthermore, the peacock's feathers do molt, so they are not an impediment year round and they are easily dislodged if grasped by a predator. Most importantly, though, is that the tail feathers act more as a governor of reproduction, rather than a "kill" switch. They do not prevent reproduction; they simply rein it in, allowing the peacock to realize his optimum number of offspring. Without the tail, peacocks would have longer lives with more years of overpopulating potential. Without the lekking behavior and the discerning peahens, peafowl could potentially breed like rabbits, with disastrous consequences. The peahens, by choosing older males with the greatest tail shaking ability and the largest, most costly implements, ensure that only the fittest gametes with proven survival genes get transmitted to the next generation. It also ensures that the peacock will not reproduce too much. By passing on the gene for the outrageous tail she ensures that her male offspring inherit the ability to efficiently reproduce without over taxing their already constrained environment. She gives them the gene for finding their magic number for efficient gene dispersion.

This notion also helps to explain why hens of dimorphic species seem to prefer older males[11]. While one could argue that old age might be a surrogate for possessing good genes, it does not necessarily translate to success in reproduction. In fact, younger cocks, with younger gametes might be more successful. Other polygynous species with less flamboyant, less potentially deadly ornaments, like throat wattles or knobs, may develop these oddities as a marker of age, not (both) as a measure of overall genetic fitness, but (and) as a marker of how likely he is to be gone after procreation. In this scenario, young males possessing genes that attempted to trick the females by simulating the aging process would be penalized for overpopulating and putting too much stress on their offsprings' survival. Again, less is indeed more.

Furthermore, it explains the immunosuppressive effect of testosterone and its relationship with the evolution of dimorphism without having to make circuitous hypotheses on the beneficence of immunosuppression.[12] More testosterone, which has been shown to be immunologically bad for the male who possesses it, is good for the female who selects for it in the male. It is also good for both male and female offspring, because the male offspring will not exhibit the phenotype until he is independent and sexually mature. In other words, the male offspring will not be impeded during development until he reaches sexual readiness.

The need for sexual selection to explain sexually disparate characteristics is an archaic and anthropomorphic notion. It relies on antiquated stereotypes and subjective, untestable notions of attractiveness. It does not adequately explain species' adaptations that do not neatly fit into its paradigm and is unnecessary. Natural selection can adequately explain the development of these sexual differences. It requires the realization that potentially malignant masculine traits can be selected for by the female, if they are adaptive for her and their offspring.

As Dawkins elaborated in *The Selfish Gene*, the evolution of sexual differences revolves around exploitation of the female egg by the male gamete. Differential size development of the male gamete to a smaller, more mobile entity was accompanied by evolution of the female gamete to a large, energy intensive (compared to the male gamete) contributor to the offspring. The strategy of the male individual reflects the strategy of the male gamete in general. The male strategy is to have little energy investment in the offspring coupled with widespread distribution of gametes or offspring. Quantity is favored over quality. The individual female strategy reflects the energy investment in creation of her gamete, that is intensive investment in a single (or small number) of gametes or offspring. (Quality not quantity) This situation is rife for exploitation by the male of the species.

The adaptive value of the male to the female and offspring after copulation is likely proportional to the amount of paternal investment in childrearing that the male exhibits. Therefore, if there is little or no paternal investment, then there is actually a survival impediment for the hen and offspring because not only is the male consuming food that might otherwise go to the hen and chick but also he is creating other mouths to feed with other females of the species. Thus it follows that the female of the species should select for a mechanism to curtail this environmental pressure and even the playing field.

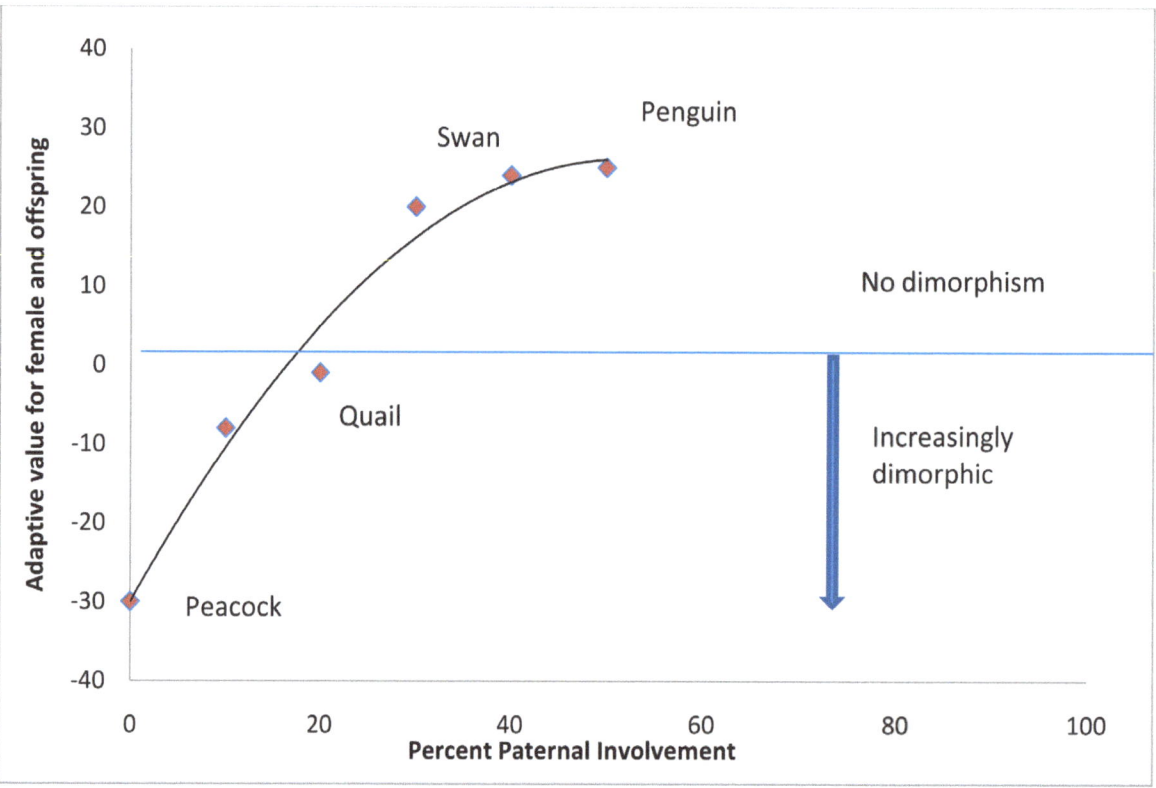

Figure 3. Plot of different species' male's paternal involvement versus their relative value towards their mate and offspring. (Values are hypothetical)Species with increasing paternal involvement in childrearing have increasing survival value for female and offspring. Those species' males who invest little or nothing to childrearing have negative adaptive value for his mate and offspring. Species above the blue line would not expect to be dimorphic. Species below the line would be increasingly dimorphic inversely proportional to the male's adaptive value to the female and offspring.

If this were hypothetically expressed, (Figure 3) then that species whose paternal investment was so low that it negatively influenced the survival of the female and offspring, some mechanism should be selected for that would combat that trend, even to the detriment of the mature male of the species. The female is turning the male's tendency to exploit the female against himself in a beautifully ironic, Ju-Jitsu-like adaptation.

Perhaps it stretches the imagination a bit, but suppose the peacock's tail is equivalent to our concept of alimony. In our mostly monogamous society, payment for the right to philander is required as well; sometimes the payment is more than would be required to ensure the well-being of the wife and children left behind who are sometimes fully grown. The jilted spouse is making it hard for the male to start a new brood lest he should decrease the emotional and financial well being of her offspring. The peahen obviously cannot extract her pound of flesh through the bar, (though she might be acquainted with a hyena or two) so she relies on decreasing the cock's fecundity in another way. Or as comedian Robin Williams once said, "Ah, yes, divorce... from the Latin word meaning to rip out a man's genitals through his wallet."

Does this line of thought apply to intrasexual selection as well? For instance does this explain why the elephant seal bull is 3-4 times the size of the cow? Or why deer species have developed elaborate antlers? On the surface, it seems that this type of dimorphism has developed to fend off rival males competing for the harems of available females. Such machinations are in place to ensure that the strongest, fittest bulls are allowed to pass on their genes. In most of these dimorphic species there is little or no paternal investment in child rearing. The females and calves might benefit if they selected for a strategy of mating that resulted in death of the male shortly after the rutting period. Indeed, elephant seal bulls do not eat for 2 months during the rut and expend valuable energy in activity directly opposing the

interests of their own survival. Males peak at 11 or 12 years of age, as reflected by dominance rank and mating success. At age 13, males are "over the hill", and prevented from entering harems. None survive to age 14. Female's life expectancy is 20-24 years.[13] Are the females driving the early death of the males to their and their offspring's benefit?

Figure 4. Male and female elephant seals, Note dimorphism in size and snout morphology. Photo courtesy of Wikipedia and National Oceanic and Atmospheric Administration, original image by Jan Roletto.

Biologists have recognized the differential life spans of males and females of dimorphic species and have attributed the decrement in male life span to the hazards of being male.[14] In other words, the male infighting, the focus on copulation and not nutrition, and the decrement in immunity due to high levels of testosterone all lead to a decline in longevity for the male from extrinsic causes. Thus if the males are dying off early from the consequences of rutting, then there are fewer older males about to pass on any longevity genes that they might possess. Absent from these observations, however, is a plausible explanation for why it is beneficial for the male to adopt and persist in using a reproductive strategy that seems so clearly mal-adaptive for members of his sex.

Some species of deer have evolved massive and intricate head gear that seem ridiculous to their function. The now extinct and poorly named Irish Elk, the largest deer known, had a massive 12 foot antler span that couldn't have been easy to negotiate even through the sparsest woodland. Famously, Stephen Jay Gould wrote an article in the journal *Evolution* in 1974 concluding that the antlers were not oversized relative to the size of the deer compared with his smaller ancestors and their antler size.

Gould argued that the antlers were perfectly sized for display and intimidation. However, this seemingly over the top rack must have cost plenty of males their life, if not diverted energy. It might make sense that such an implement might be selected for by the female because it allowed the offspring to have more forage food by killing off all but the absolutely fittest males. There would be few old codger elks banging these antlers around in the woods eating up food that might be better served for his offspring.

Figure 5. Irish Elk, (Megaloceros giganteus), skeleton. From Wikipedia, photo by Franco Atirador.

This arrangement of harems and dimorphic males would seem to have been arranged by the male of the species in question. After all, he gets to have lots of sex and lots of offspring so it must have been he that "decided" on this arrangement. Or in other words, selection pressure for increased size or antler length occurred independently of female selection pressures. However, short of coercion, the female is the one that decides when and if copulation is to occur; might it be more reasonable to assert that it is the female who is driving the establishment of this arrangement to her benefit? She gets the fittest gametes from the male from the gladiatorial arrangement; the male, who has evolved to be superfluous after copulation anyway, doesn't last long once maturity is reached. He has expended so much of his energy into fortifying, fighting and copulating[15] that he is denied his own longevity. She doesn't have to divert energy from survival and child rearing into being attractive. Thus perhaps it is not the male herding his harem; it's the females "herding" and shaping the males to preserve their stable strategy for species survival. The male assents to the arrangement because in dying relatively early, more of his offspring survive.

This also has some interesting resonations as to what the female of a species finds attractive. "Dangerous" traits are attractive for the female of species where the paternal input on child rearing is absent. Conversely, penguin hens select mates who are fatter, with lower vocalizations (indicating more fat in the larynx), and are more likely to survive the long brooding period; swans, who are monogamous for life, evoke rigorous courtship displays that are likely to reveal the stamina and parental fitness of one another; these hens, who need the father around for chick rearing, do not value gaudiness.

Is this the reason the female of our species finds "dangerous" men attractive? Though *Homo sapiens* is only moderately dimorphic, there is a definite difference between our sexes. Though current paternal input in childrearing is comparatively high in our species compared to other dimorphic species, there is likely to have been a certain amount of polygamy. If the female and offspring, during resource constraint, were better off without so many polygamous and prolific males about, then "dangerous" traits like aggression or risk taking behavior might be advantageous for the female and offspring and be naturally selected for by females. This might be especially true in matrilineal societies where brothers, not husbands, look after their sister's children. Then again, sometimes it's difficult to try accounting for taste.

This idea has important implications for the delineation of evolved traits. Evolution by natural selection, belied by it succinct and beautiful premise, is far from a simple process. There is an ongoing tug of war between the forces involved in the betterment of individuals to avoid death and the development of seemingly detrimental adaptations that ensure a gene does not get too big for its collective britches.

CHAPTER 4

ALTRUISM

Arose by any other name, might smell as sweet; altruism, though, by another name, is slow death. Under the basest scrutiny, individuals that truly act unselfishly are asking to die prematurely, either via potential starvation or increased risk of predation. What an individual gains from unselfish behavior may be beneficial in the long run; but it might not. It's a huge leap of faith to assume I'll be paid back for my unselfish act today, if by doing so I might not even make it to the day of reckoning. Individuals with a tendency toward unselfish behavior might find that they and their offspring are living longer. This is the crux of the argument of the gene centrists' view of altruism. But by exhibiting selfless behavior, individuals are setting themselves up to be exploited by selfish individuals, by the very definition of altruism, and their survival should be lessened. How can altruism start amongst a herd of selfish animals?

In trying to fathom the necessity of death under natural selection, we must also scrutinize altruism. Altruism and death go hand in hand. It could be argued, that as behaviors (if death can be regarded as a behavior; I would argue that it can), death and selfless behavior are simply on the same spectrum of altruistic acts. Death might be rightly regarded as the ultimate act of altruism.

It could also be argued whether or not there is truly altruistic behavior present in the animal kingdom. Any seemingly selfless act could be interpreted as behooving one's genetic heritage (kin selection) or mandating reciprocity (reciprocal altruism). An animal like the prairie dog who takes his turn standing watch while others forage is reducing his forage time and fitness temporarily, but he expects to be repaid by others in the group when it's not his turn to stand guard. The benefits outweigh the temporary detriment and as such could be viewed as cooperative selfishness.

The peacock's tail is, one could argue, a form of altruism, in that the cock exhibits traits that decrease his fitness to the benefit of the hen and the brood. This mechanism for the evolution of the peacock's tail is a theory, yet to be validated. But it opens the door to the idea that subtle adaptations can arise that decrease an individual's fitness driven by his species' mates. We have seen how a female peacock can in theory select for traits in the male to her and her offspring's benefit. How then could a trait like altruism develop and be selected for without relying on a group selection mechanism?

To understand this we must understand dominant and recessive genes. One must realize that a gene can exist in an individual without being expressed. Everyone has two copies of a particular gene, one copy inherited from each parent. Typically, one form of the gene will be dominant and if inherited will determine the trait in that individual no matter what the other gene has to say. The gene for brown eyes, for example, is dominant over blue eyes; the only way one gets blue eyes is if a person inherits two copies of the blue eye color gene. In this instance, blue eyes are termed recessive. Sometimes the recessive gene will have an impact on the trait at question, causing a blending of the traits, say hazel eyes, in our above example. The gene for brown eyes would be incompletely dominant in this instance.

The expression of genetic traits is not always this simple. Complex behavior like altruism is not likely to be controlled by such simple inheritance patterns. Genes or groups of genes acting in concert can have complicated patterns of inheritance. Nor should we deny that there may be some learned aspect of behavior that is not under genetic influence. However, what is important is that there are likely to be genetic tendencies towards certain behaviors whose level of expression *is* under genetic control. The expression of the trait may not be like an on/off genetic light switch; it might be more like a genetic rheostat where varying degrees of expression can be exhibited and may be alterable depending upon environmental conditions.

With that as a background, perhaps we can understand how a gene for altruism could be selected for. Let's assume that there is an altruistic trait that unequivocally leads to a decrease in the reproductive fitness of the individual that inherits it. He receives no benefit later on in reciprocism. I postulate that altruism is a gene that is deadly or at least leads to a decrease in fitness for the individual that inherits the full complement of recessive genes, but beneficial for those who inherit only a partial, and thus incompletely expressed, complement. To better explain altruism's mechanism of persistence, let's turn our attention to a decidedly deadly gene like Tay-Sachs (TS) disease.

TS is an inborn error of metabolism that in its homozygous recessive form is universally fatal by age four. In other words, it requires two copies of the gene from both mom and dad, to inherit the disease. Other diseases with similar lethality and gene expression like cystic fibrosis or sickle-cell disease, have arisen, it is postulated, because there is some benefit to having only one copy of the gene. People with heterozygous (Ss) genotypes, i.e. with only one copy of the gene for sickle-cell disease, are relatively normal in physiology but are slightly protected against the ravages of malaria. People with one copy of the cystic fibrosis gene are thought to be relatively immune from the effects of cholera. Thus it is easy to see how these genes would be widespread in areas where these diseases are endemic. However, there is no rigorously known benefit for the heterozygosity of diseases like Tay Sachs, Niemann-Pick, and a whole host of other genetic mitochondrial defects that are lethal even before the child reaches the age where he or she might be able to pass the trait on. Most biologists assert that Tay Sachs and other genetic diseases persist through genetic drift throughout a population that is either isolated or culturally inbred. Detrimental mutations can persist because natural selection with its tendency to weed out malignant mutations is weaker in populations that are sheltered from the input of the population as a whole which will likely contain more beneficial genes. High on an isolated mountaintop, for, instance, "bad" genes can proliferate because the "good" genes that could outcompete the bad ones aren't as plentiful. There is debate on the relevance of drift and the founder effect as a mechanism for the frequency of certain alleles in a population only because the genes in question are seen as maladaptive. What if these malignant genes could be seen, under certain conditions, to be advantageous? If only a slight advantage, the so called "malignant gene" would be selected for.

To see how a seemingly malevolent gene might be advantageous, imagine an enclave of a small number of animals where the frequency, for simplicity's sake, of a malevolent gene (m) is 50%. All homozygotes (mm) with the malevolent gene die at the first sign of food scarcity. Heterozygotes (Mm) have no impact on their longevity. Genotypes MM (25%), Mm (50%) and mm (25%) are possible. Imagine a subpopulation of 4 individuals of this population who face a drought. They only have 8 units of sustenance available from the environment during this time but need 3 units each to survive this period. If the food were randomly distributed, these are the possibilities (see the table). One animal might get all the food, leaving none for the rest. It might be evenly distributed among 2 with one getting none. Most likely it would be evenly distributed but all cases are listed. Because one animal has the homozygous malevolent gene he dies and requires none.

A gets...units	B gets...units	C gets ...units	D (mm) no units	# Surviving
8	0	0	X	1
7	1	0	X	1
6	2	0	X	1
6	1	1	X	1
5	3	0	X	2
5	2	1	X	1
4	4	0	X	2
4	3	1	X	2
4	2	2	X	1
3	3	2	X	2

Because the gene frequency of the malignant gene is 50 %, twenty-five percent of the population would be MM, 25 % would be mm and 50 % would be Mm. Two of the three potential survivors above would be expected to be Mm and one would be MM. There are 10 scenarios of food distribution, with a 40 % (4/10) chance of 2 animals surviving the drought. Because statistically 2 of the three animals are Mm and 1 is MM genotype, when 1 animal survives there is a 66% chance it is an Mm animal; when 2 survive it is 100% probability to include an Mm animal. Thus the probability of an Mm animal surviving the drought is at least 80%. { [(.66) x 6 (no of 1 surv. scenarios) + 1.00 x 4 (no of 2 surv. scenarios)]/ 10 =.8}

The same example using 4 animals in a population with the same reproductive drive without the malignant gene (all MM) shows 14 possible scenarios as listed below. Remember there are 8 units of sustenance and each animal requires 3 units to survive the drought.

One gets…	The next gets…	The next gets…	The last gets…	# Survivors
8	0	0	0	1
7	1	0	0	1
6	2	0	0	1
6	1	1	0	1
5	3	0	0	2
5	2	1	0	1
5	1	1	1	1
4	4	0	0	2
4	3	1	0	2
4	2	2	0	1
4	2	1	1	1
3	3	2	0	2
3	3	1	1	2
3	2	2	1	1
2	2	2	2	0

There are 5/14 scenarios where 2 animals survive (36%) and one scenario (7%) where none survive. Thus in this simplistic example, the population with the malevolent gene is 4 % more likely to keep 2 animals alive (36% without the gene vs. 40% with the gene). Furthermore, in 7 % of scenarios (1/14) for the population without the malevolent gene no animals survive. And when food is most evenly parceled (the last row of each table and arguably the most likely scenario in nature) 50 % (2 of 4) survive with the malevolent gene in the population and none survive in the population fortunate(?) enough to not have this deadly gene.

One could argue this is simply a complicated way to state the obvious: fewer mouths to feed equates to more survival during resource constraint. And it might be argued that this is too simplistic to be a reflection of the natural world. But it highlights the quirky nature of heterozygous gene expression in an environment where there are more mouths than meals. The inheritance of a gene that only has the potential to kill if it is homozygously expressed, has a counterintuitive benefit for those in whom it is not completely expressed.

It also reflects the tendency of organisms to push the boundaries of what the environment can sustain. In any ecosystem, the tendency for consumers is to expand population numbers until the limit of what the environment can sustain is reached and then push just a bit more for good measure. There will always be more mouths to feed than there is feed available for all the mouths. The trick, for a crafty gene, is to find the balance between gene number and gene survivability.

Now, finally, if one substitutes an altruistic gene for the malevolent gene and ascribes a phenotype that acts with unselfish concern for the well being of others to his own detriment, it is easy to see how altruism might take hold within a population. Though the trait of altruism may not seem as malignant as a genetic defect like Tay Sachs disease, this concept substantiates the notion that genes that appear individually detrimental can actually be beneficial for gene dispersion. Just as the heterozygous expression (Mm) of the malignant gene increased after one generation in the above example, so could the prevalence of an altruistic allele increase over many generations even if the full expression of the altruistic trait was not advantageous. It permits the allele for altruism to spread widely even if the homozygously (aa) expressed trait does not. It allows individuals with the heterozygous trait (Aa) to out compete those who live in populations without the gene. It appears that individuals in groups with the trait out compete those individuals in groups without it, even though it need not invoke a group selection mechanism.

In a similar way, the peahen exploits the peacock by selecting for an ever more ridiculous impediment and thus reduces environmental pressure favoring her and her offspring, the gene for altruism is selected for because it subtly and silently reduces environmental stress for the individuals among populations where the gene is present. Although the peahen seems to be selecting for the cock's tail, really it is nature which is selecting for the tail, in that natural selection favors the peahens (and her offspring) who choose peacock's with larger and larger tails. In the same way, the gene for altruism spreads throughout a population because individuals with only one copy benefit by the exploitation of individuals with two copies of the gene. Altruism has a doubly beneficial effect for populations with the gene. Not only is there benefit to others in the population from the altruistic acts of the homozygote altruist (sharing food, crying out for predators), but by being less fit, he eases population pressure when he dies. The heterozygotes are exploiting the homozygotes, allowing the gene to spread. Or more paradoxically, the gene is exploiting itself. It seems there may be some truth to the expression that only the good die young.

Additionally, altruism need not be homozygously expressed to have some effect (Nor does it require Mendelian patterns of expression; simply having differing degrees of expression of the trait under influence of one or more genes would suffice). Heterozygotes could have some altruistic tendencies depending upon environmental conditions. These conditional altruists could turn it on or off depending on the scarcity of resources. Thus, the heterozygotes, in times of stress, would not be penalized for their altruistic ways (because they could turn it off), but would benefit in ways that are conventionally postulated for the benefits of altruism (reciprocation, game theory) in times of plenty. In the same way that the CF or sickle-cell gene spreads because there is a benefit to its heterozygous condition and little penalty, so could the gene for altruism.

Does this mean that altruists are genetic pawns to non-altruists? In other words, like sickle cell disease, the homozygous condition would be decidedly detrimental. Pure (aa) altruists would have reduced fitness compared to non-altruists (AA) or partial altruists (Aa). The gene spreads throughout a population because of the increased fitness of the heterozygotes who benefit from their altruistic brethren or neighbors. If one thinks of sickle cell disease patients as genetic pawns to those fighting malaria, I suppose it could be argued that altruists are pawns to those fighting starvation and predation.

What does the homozygote get out of the relationship?

Well, to be frank, he gets screwed. But the gene for the behavior can flourish because it is successfully and silently carried by brethren who benefit from its full expression in others. It's a tough pill to swallow for a morally conscious natural selectionist, but nobody ever claimed nature has a morality clause. Most would believe that there must be some benefit to survival for the individual with any long lived, expressed trait like altruism. This need not be the case so long as the net overall effect of a gene is positive. Just as the homozygote with a deadly genetic disease gets a bum deal, so does the altruist; however, the other individuals in the population can outcompete others outside the population without the altruistic allele.

The conclusion that there may be a penalty for being nice may seem disheartening to some. But there also may be a reward. Once a critical mass of altruists has been produced, the survival benefits may outweigh the detriments of the self-less behavior. The altruist may gain status, as Professor Zahavi has pointed out. A by-product of this generosity may lead to offspring and gene longevity. The altruist wraps himself into the fabric of his species, but the origin of the behavior is nestled in the recessive gene.

Thus the gene for altruism is helping a population find its magic number, not through group selection mechanisms, but by individual selection for genes that increase the overall fitness of one's progeny.

This carries some not so intuitive consequences. What is "good" for a gene is not necessarily "good" for the individual carrying that gene. Altruism, for example, is not immediately adaptive for the individual carrying two copies of the gene. Sharing food or calling out when predators are near puts oneself at risk and should lessen the probability of one's own survival. But so long as the conditional altruists (i.e. the heterozygotes, Aa) can survive until maturity and can confer the altruistic phenotype to some of their offspring, it is adaptive (beneficial) for the gene itself. More individuals in the brood with the allele will be more likely to survive than those individuals in broods without it. The gene efficiently spreads in populations where resources are constrained.

It is difficult to envisage a scenario where increasing the likelihood of death would be inherently adaptive for an individual. In other words, having a gene that results in one's own death can hardly be good for dissemination of that gene. If, however, death of the individual meant increased fitness for three quarters of his brethren, then that gene would be selected for so long as broods or individuals in populations without the gene had less survival.

What about organisms that don't have dominant and recessive genes or sexually reproduce but still show features of altruism or group level selection? For instance, the asexual bacterium, *Pseudomonas flourescens,* when placed in a liquid medium quickly proliferates and uses up all the oxygen present in the medium whence death ensues for all. Some of the bacteria, having mutated slightly, will then start secreting a buoyant protein that allows all the bacteria (whether they have helped make the protein or not) to survive on the surface and float where they can get to the oxygen from the air. Because the "freeloading" bacteria are not using energy towards making the protein, they can reproduce more efficiently and outnumber the altruistic subtypes. The mats get laden with the extra mass and sink thus dooming all the bacteria. The argument has been made that those mats that enjoy prolonged survival are composed of bacteria that are subverting their own selfish procreational needs for the good of the colony or group.[16]

Or could there be another explanation? Suppose a mat was composed of organisms that all possessed the gene to produce the protein for floating. If the incipient mat were subsequently colonized by non-producers, the mat would quickly sink, before it had a chance to be established. However, what if a mutation developed in a protein-producing bacterium that retained the gene to make the buoyant protein but turned off its production when oxygen levels were high? These conditionalists would appear phenotypically different from the obligate producers when they were plated out and speciated under normoxic conditions. The conditional producers would not make the protein when grown on plates under normal oxygen concentrations and their colonies would appear different than the obligate protein producing strain. These bacteria would have a reproductive advantage over their cousins who still continuously produced the protein. But at the genetic level, the gene encoding for the protein would still be present and would not be penalized for always diverting energy to actually making the protein. When the mat becomes heavy and oxygen levels drop, the restriction would be lifted and more protein could be made by the conditionalists to keep the mat afloat. The mats that persist would be composed mostly of bacteria that have the ability to make the buoyant protein, but when plated out and speciated under normal oxygen conditions, it would appear that one strain (the conditionalists) would not be making the buoyant protein. Natural selection at the level of the gene for the buoyant protein doesn't discriminate between the two strains because both have the gene for making the buoyant protein. The conditional protein producers seem to be exploiting the obligate producers. But in actuality, the gene, with the help of the "on/off" switch, is finding a way to exploit itself to become more widespread.

The above scenario is simply a hypothesis; it, in all likelihood, is not what is going on in these mats. It simply illustrates a mechanism that a gene could employ to be able to exploit itself. The gene can be spread widely, even if the trait that the gene codes for does not.

This idea of gene self-exploitation can also explain how reciprocal altruism can get started in a population. Vampire bats for instance have developed a complex system of blood sharing. If a bat goes without eating a blood meal for two days in a row he will die. Bats have developed a system of regurgitation, such that those who are unsuccessful can approach those that have been successful and bum a meal from the successful bat. Researchers have shown that those that are more closely related are more likely to share blood meals with one another. If they are not related, the bats will establish a tit-for-tat strategy; that is they will only regurgitate to those who have regurgitated to them in the past. One problem that theory of reciprocal altruism has in explaining these altruistic acts between unrelated individuals is that the gene for the behavior handicaps the donor. It works only if the cost of the donation is less than the benefit he might receive later. The donor needs a crystal ball to know if his altruistic actions today will bear fruit later on.

Once a relationship is in place, the behavior pattern makes sense. But how does it get started? It's a leap of faith for the first altruist amidst a host of selfish bats to assume that it will be repaid later on. Does the gene for the behavior arise simultaneously in unrelated individuals? Perhaps it evolves from kin behavior, but how does it move to non-kin behavior?

It might be easier to explain if the trait arose as a recessive gene or conditionally expressed gene. An animal with two copies of the gene expressed altruistic tendencies and initiated the sharing of a meal. It runs the risk of being exploited and sometimes is duly penalized with decreased fitness and dies younger than would be expected without the gene. But the bat will only take his generosity so far. If the recipient bat does not reciprocate later on, then it won't continue to donate to him. It is the tendency to initiate meal sharing, to take a chance on being exploited, which is expressed in the recessive phenotype. The heterozygous animals with only one copy (or those who could repress the complete expression of the gene or genes) would substantially benefit from having the altruistic homozygous phenotype (aa) in the gene pool but would not actually be penalized for only carrying the gene. These animals could outcompete those bats in other populations where the gene doesn't exist. Additionally, vampire bats are likely to have a limited number of reliable food sources. Overpopulation could be an issue. By having animals with reduced fitness around, the altruists sacrifice not only their food but also ultimately they sacrifice themselves, all for the good of their genes which have been safely and silently ensconced in their surviving heterozygous brethren.

Furthermore, if the effects of any seemingly maladaptive gene (like altruism) were delayed enough for the gene to be passed on to one's offspring, then it would be beneficial for the gene, if the parent were to die. Once the gene is safely ensconced in the independent offspring, the parent's increased probability of death becomes adaptive for the gene, because it is adaptive for the offspring. Without the "killer" gene, the independent offspring might be forced to compete for food with the parents. If nobody ever died, overpopulation would quickly ensue, resources would vanish and all would perish. Also, it augments a family's efficient reproduction and survival. Those with the gene that allows regulation of population numbers at both ends of the birth to death spectrum will be more efficient than those with only the ability to control birth rates. The gene becomes wide spread because individuals with the gene can out compete those without it.

And also why altruism really doesn't appear phenotypically until the individual has reached sexual maturity. We say someone is mature when they reach sexual readiness, but also emotionally when they begin to exhibit selfless behavior. Dependents of any species, including *Homo sapiens*, are very selfish. It is not until they are ready to pass on their genetic material does it make evolutionary sense to exhibit altruistic behavior. In other words, decreasing one's fitness isn't a sound practice unless you're ready to pass on your genetic material.

Until now we have examined altruism mostly in non human species. Though we should examine it in humans, we should do so cautiously because it becomes prickly to suggest that human behavior is strictly gene centered. Much of our culture is learned and socially cemented. We are socialized to altruism by group mores. We are pushed from just past infancy to act unselfishly; status can be determined or at least influenced by how much we give away. It's no coincidence that the Board of Directors is chosen from the biggest benefactors. Yet, there are those rare people who do act selflessly without secondary gain and risk their lives to do so. They seem to be instinctually selfless. Joyce Roush, who was one of the first persons in the United States to give away one of her kidneys to a complete stranger, is an example. Those instinctually altruistic would have the equivalent of two copies of an altruistic gene. The remainder would have the equivalent of one or no copies of an altruistic gene and would be mostly selfish or altruistic only under certain circumstances.

Any instinctual behavior is thought to be genetically driven, not something learned, nor reasoned. It is a tendency to act in a way that is part of one's make up, part of who you are. Some are impetuous. Some are manic or some shy. Some are greedy and some not. It is the behavior that emerges when you don't have time in a situation to think about the outcome. Lending strength to this notion is the fact that researchers are discovering that complicated behavior like altruism can be linked to the affinity of certain neuroreceptors for certain neurotransmitters, like oxytocin, in certain areas of the brain, perhaps under the influence of a limited number of genes. The best example of such instinctual human behavior that I can present you with is the story of Petty Officer 2nd Class Michael A. Monsoor.

Petty Officer Monsoor found himself in insurgent stronghold of Ar Ramadi, Iraq on September 29, 2006. I think his citation for our country's highest award for valor, The Congressional Medal of Honor, explains his actions far better than I can condense.[17]

Citation

For conspicuous gallantry and intrepidity at the risk of his life above and beyond the call of duty as automatic weapons gunner for Naval Special Warfare Task Group Arabian Peninsula, in support of Operation IRAQI FREEDOM on 29 September 2006. As a member of a combined SEAL and Iraqi Army Sniper Overwatch Element, tasked with providing early warning and stand-off protection from a rooftop in an insurgent held sector of Ar Ramadi, Iraq, Petty Officer Monsoor distinguished himself by his exceptional bravery in the face of grave danger. In the early morning, insurgents prepared to execute a coordinated attack by reconnoitering the area around the element's position. Element snipers thwarted the enemy's initial attempt by eliminating two insurgents. The enemy continued to assault the element, engaging them with a rocket-propelled grenade and small arms fire. As enemy activity increased, Petty Officer Monsoor took position with his machine gun between two teammates on an outcropping of the roof. While the SEALs vigilantly watched for enemy activity, an insurgent threw a hand grenade from an unseen location, which bounced off Petty Officer Monsoor's chest and landed in front of him. Although only he could have escaped the blast, Petty Officer Monsoor chose instead to protect his teammates. Instantly and without regard for his own safety, he threw himself onto the grenade to absorb the force of the explosion with his body, saving the lives of his two teammates. By his undaunted courage, fighting spirit, and unwavering devotion to duty in the face of certain death, Petty Officer Monsoor gallantly gave his life for his country, thereby reflecting great credit upon himself and upholding the highest traditions of the United States Naval Service.

Figure 6. CMH Award winner Petty Officer 2 nd Class Michael A. Monsoor. Photo courtesy of Congressional Medal of Honor Society and United States Navy.

Such behavior cannot be explained by a selfish gene or reciprocation later in life. It could be argued that this type of behavior could arise as a surrogate type of kin selection, where P.O. Monsoor was treating his comrades as surrogate "brothers" and was acting for their benefit. It's this same impulse to assign kinship to those close by that cuckoos exploit, to trick another bird into feeding and raising their chicks.[18] And certainly induction and indoctrination into any military unit would tend to foster the notion of kinship and brotherhood. But the tendency to act this heroically is hard to explain; every hard-fought, fairly- won Darwinian instinct should scream out to flee such a circumstance. Yet, "Although only he could have escaped the blast, Petty Officer Monsoor chose instead to protect his teammates. Instantly and without regard for his own safety, he threw himself onto the grenade to absorb the force of the explosion with his body, saving the lives of his two teammates." The instinctual, split second decision made by P.O. Monsoor can best be explained because he, like Joyce Roush and others, was an instinctual, full throated altruist.

Altruism, dominant and recessive genetic disease, and death from intrinsic causes persist because they help individuals, forming a breeding pair, existing within a population, find their genetic magic number (and thus the magic number of offspring and, collectively, of the population). It efficiently maximizes their survival over individuals in populations who don't have these seemingly maladaptive genes. It is especially important for species that have little or no predation. Some of these seemingly maladaptive genes don't kick in until after fecundity; thus it would seem natural selection would not have a strong effect on the elimination of these genes from the gene pool. However, I would argue there is some selection in **favor** of these genes because they allow populations to efficiently find their capacity in an ever changing environment. In other words, organisms with genes that result in their own death after cessation of reproduction and offspring independence[19] will be more efficient at keeping their offspring alive, especially in a resource constrained environment. Furthermore, even if the gene results in death before fecundity, like Tay Sachs disease, we have seen that there may be some benefit to individuals of a breeding pair (who make up a population), if during resource constraint, there are simply fewer mouths to feed.

It sounds incredibly cruel from the vantage point of our calorically infused society. Perhaps it is all too Kantian, that there exist traits that are seemingly mal-adaptive for an individual but are categorically beneficial for the rest of the group. However these mechanisms also allow individuals in a species to continue to have a strong desire to procreate, but not penalize the subsequent generation with overcrowding and environmental constraint.

This concept seems far-fetched that such a complicated behavior as altruism could be likened to a recessive genetic disease. Yet researchers are finding that selfless behavior is influenced by a hormone, oxytocin, which can be given intranasally. One study found that those administered oxytocin were 80% more generous than control subjects in giving away house money. (Zak et al.) Others have found that variations in the gene for the oxytocin receptor protein correlate well with how generous subjects are in sharing gifted money with anonymous recipients. (Israel et al.) How interesting is that! It seems that complicated behavior like altruism can be influenced by neuroreceptor/ neurotransmitter activity under genetic control. And what a simple way to confer a wide spectrum of behavior. By varying receptor number, receptor affinity, speed of synthesis of the neurotransmitter, speed of release of neurotransmitter, or speed of neurotransmitter removal from the synapse, an almost infinite number of responses can be achieved from the same stimulus.

CHAPTER 5

SPECIES STABILITY: THE PROBLEM OF "LIVING FOSSILS" WHEN GOOD IS BAD

In 1938, a young museum curator in the town of East London, South Africa, Marjorie Courtenay Latimer discovered an unusual fossil, though at the time she didn't fully realize the significance of her find. She hadn't dug the specimen from the ground, nor found it in a river bank or rock cut. She found it at the bottom of a pile of fish that a local fisherman had brought to port. Though she didn't recognize the exact nature of her find, she was brilliant enough to realize that this fish was different and that it required further study.

She sketched the fish, mounted it and threw away the entrails and skeleton. She wrote a local ichthyologist at Rhodes University in Grahamstown, Dr. J.L.B. Smith, some 50 miles away, but because of the timing of the find, the expert was away on Christmas break. It wasn't until over a month later that Professor Smith was able to authenticate in person the find as a living coelacanth, a species of primitive fish in existence over 65 million years ago. The present specimen seemed to have been preserved unchanged from the Late Cretaceous period when dinosaurs like *Tyrannosaurus rex* and velociraptors roamed the Earth. Imagine the uproar if a land dinosaur were to be uncovered in the jungles of some remote land.

A frantic search ensued at the highest levels of the scientific communities to find an intact specimen. Rewards were offered, dead or alive, for the mysterious fish. Fourteen years later, a second specimen was found and preserved and was able to be examined by the experts in the field. Their resemblance to fossilized specimens is uncanny. They have the same primitive spinal chord and hollow spines on the fins that give the animal its name. (Coelacanth, Gr. for "hollow spine") Its characteristic primitive tail, so different from modern fish was the feature that cemented the identification and placement of the modern species in the same class as the ancient varieties. Today these animals have even been filmed by deep diving oceanographic submersibles.

Figure 7 Marjorie Latimer and her find. Photo courtesy of Dinofish.com website.

Creationists have long celebrated the discovery of the coelacanth as a fly in the ointment to Darwin's theory of evolution. They crow about the stability of the species that hasn't changed since time immortal. A follower of Darwin would predict that over the period of time that the coelacanth has enjoyed, we would expect to see an accumulation of mutations to make one's distant ancestor unrecognizable from current examples. There should be mutation's that increase a species survivability and reproductive success.

Most would counter this argument stating that for some species there is no pressure from the environment to induce change. The species is saying, "Let's leave well enough alone. We have found a stable body form and foraging strategy and we don't need to change."

In environments like the deep sea of the coelacanth, where sunlight levels do not significantly vary, there is also likely to be stability of resources. Food availability will be consistently under abundant. There is no pressure to change in this scenario. In fact there is likely significant pressure to stay the same. Any change in reproductive success is potentially bad. Mutations that lead to a trend towards increasing the number of those surviving to fecundity could lead to a decrease in the number of those actually surviving. An environment that could support four fish, if pushed to support eight, might result in all eight of those with improved reproductive success dying out. There is actually pressure to not improve one's reproductive lot in life and live happily ever after without changing too much.

It could be argued that any change in reproductive success would not likely be this dramatic. A mutation could occur that would subtly effect the survival of offspring that would result, for example, in an animal developing a protective mechanism to ward of predation. These animals would not add to the species number but would outcompete and replace those without the mutation. And for any forward thinking ethologist this makes intuitive sense and is the backbone of Darwin's concept of natural selection. But in the near term, it may not be wise for an individual to boost his reproductive success if the environment cannot support the achievement. Those with the mutation must compete with those with a stable environmental and reproductive strategy. They may be able to ward off predators more effectively and have more offspring survive but they are no more effective at foraging than their unmutated counterparts. Now, though, there are more mouths to feed. At first the mutated animals would be able to outcompete their unmutated brethren and the population numbers of the mutated fish would increase. But at some point their numbers would exceed the carrying capacity of the environment that had been so painstakingly established by their unchanged relatives. Areas where the mutated fish were abundant would be devoid of nutrients, if they were able to rapidly expand their numbers. Throw in a periodic time of relative environmental dearth (like a la Niña or volcanic ash cloud) and you have a recipe for extinction of the mutated individuals. Furthermore, any increase in the survivability of the parent may have a decremental impact in the survival of the offspring through increases in resource scarcity. Their "improvement" under certain circumstances is actually a detriment. Because the unmutated animals have developed a stable strategy with predictable predation, there is more food to go around for the survivors, especially the offspring. Just as the hare cannot keep up his torrid pace in the long race against the tortoise, the animal that improves his reproductive success too rapidly may lose out to the animal that keeps his environmental strategy slow and steady.[20]

Newton stated that in the physical world, for every action there is an equal and opposite reaction. In physiology, organisms regulate pH and temperature by counteracting any changes from the optimum number. We should expect that natural selection should work in a similar fashion once a species' number has reached its equilibrium with environmental resources. In a resource fixed ecosystem, individual adaptations that foster increased progeny would have to be countered by a decrement in either longevity or an improvement in individual resource acquisition in order to maintain a viable gene number. Individuals with increased progeny run the risk of being outcompeted by individuals who keep numbers at the carrying capacity of the environment. In resource stable environments like the deep sea or tropical rainforests, it is thought that species stability is accounted for by lack of selection pressure to change. It seems that there might be selection pressure to remain the same.

The coelacanth is not immune to genetic change. In fact today's coelacanth gives birth to live off-spring, presumably, something its ancestors did not. Today's animal is classified as a different species than its ancestors. Thus there are selective pressures in place that are coaxing the coelacanth to change. The lack of *significant* change should be attributed to a selection force counteracting the mutational coercions resulting in evolution. Simply stating that the coelacanth remains unchanged because there is no pressure for the coelacanth *to* change is not sufficient. Our roiling nucleic acids are pleuripotent engines for genetic change and provide a constant, buoyant force for evolution to occur given enough time. If there were no counteracting downward force on this metaphoric balloon, it would change its course. If there were no selection pressures keeping the primitive fish in its ancestral condition, it would change over time. Thus natural selection does not always result in evolution. Natural selection can result in species stability.

The idea that natural selection can result in species stability might seem to some Darwinists to be a very flimsy argument. However, as the lawyers say, the existence of the modern day coelacanth is a thing which speaks for itself. Millennia of stability speak to the selection of the stable coelacanth form over its potentially mutated, extinct brethren. There must exist organisms, under certain conditions mimicking those of the coelacanth, which have reached such sublime equilibrium with their surroundings, where any change, seemingly good or bad, would have been detrimental to the survival of offspring. One cannot discount the impact a seemingly beneficial mutation might have on the survival of an individual's offspring, when one's own improved survival negatively impacts one's offspring's survival. If we revisit the salmon, what if a mutant salmon didn't die after reproduction? That would seem to be a beneficial mutation for the individual; but if that mutation were to spread, it might be seriously detrimental for the offspring of the mutant salmon in overcrowding and resource constraint. Trying to cheat death can have disastrous consequences.

The argument ultimately reverts to semantics. Darwin's theory of natural selection rests on the most impervious layer of scientific bedrock, anchored in place by strata upon strata of observation and analysis. The conundrums arise when we try to interpret which adaptations are beneficial for an organism, breeding pair or population. Our poor powers of reasoning often can't account for the mysterious ways seemingly harmful adaptations can positively affect the survival of an animal's genetic heritage; the obverse should also be true as well. The interpretation of seemingly beneficial adaptations can have pitfalls as well. Perhaps, under certain conditions, t'is folly to be wise, or at least naturally selective, when stability is bliss.

Chapter 6

CONTROLLING OUR MAGIC GENE NUMBER:
Reproduction and Death

To survive and prosper in this dog eat cat world, organisms have developed a number of strategies along a spectrum of reproductive and rearing possibilities. There are some organisms who still essentially clone themselves asexually, like bacteria, or some who swing both ways, reproducing both sexually and asexually. However, most organisms have developed a sexual reproductive strategy; that is the genetic material of two organisms is combined to create offspring with two sets of chromosomes or two copies of the DNA blueprint of that species. This provides redundancy for the complicated genetic code of higher animals and allows genetic material to mix, match and scramble. This sounds unfavorable, but it is the driving force of variability and thus the main motor of evolutionary change. It is why you don't look exactly like your mom or dad (but doesn't explain why you look like the mailman.)

The explanation of why most species employ sexual reproduction instead of cloning has been difficult for the gene centrists; if it behooves a gene to be widespread and self-centered, then why does it employ sexual reproduction with its 50-50 risk of not passing on the gene in question? Is there not a reproductive penalty for needing two to tango? What if you can't find another of the opposite sex? This two-fold cost of sex has left many gene centrists in the red.

For one, if we look back at our "efficient" gene hypothesis, it may be better to be incapable of reproducing all the time. By evolving towards sexual selection with its constrained ability to produce offspring, organisms might have found they could survive better without so much competition from sibling clones.

More importantly, sexual reproduction allows the recessive gene to emerge as a potent force for natural selection. Sexual reproduction alone can rapidly disseminate recessive alleles compared with asexual reproduction. In both types of reproduction, having two copies of genes allows a proportion of genes to be quiescent; it allows more genetic substrate for variance to occur without penalizing the bearer if the variance goes astray. It allows natural selection to tinker with the recessive gene with no penalty for malfeasance. Just as an artist has colors on his palette ready to use and mix, so does our genetic blueprint. By having quiet recessive genes about in the pool we don't have to fabricate a new gene from scratch to achieve a successful adaptation. Like a painter, we don't have to grind and process the solid pigments when we want a new color. We already have the paint on the palette, ready for a slight alteration if we want a new color. If we don't like the color, it need not always be expressed on the canvas. If an unfavorable mutation occurs in a recessive gene, it is unlikely to affect the survival of the carrier; but unlike asexual reproduction, if the recessive mutation is helpful, it will spread only in the organism that reproduces sexually. (The only way a recessive gene can establish itself in asexual organisms is for the mutation to occur twice.) The recessive gene is the research and development laboratory of the sexually reproducing genome. What a powerful tool to have in the tool box. It is easy to see why most species utilize this form of reproduction.

Furthermore, it allows even genes which are lethal like the CF and sickle-cell alleles to be adaptive for those with only one copy and relatively widespread in the population. Having two copies of genes doubles the amount of variation possible; having recessive and dominant aspects of sexual inheritance allows experimentation by nature on half our genes without penalizing the carrier if the outcome of the experimentation is not beneficial. Most importantly, sexual reproduction allows dissemination when it is beneficial.

Once organisms had decided to reproduce sexually, they then needed to work out a strategy for making it work. On one end of the spectrum, animals like fish or amphibians rely on generating lots of offspring and hope a few survive. They have a multitude of offspring where parental involvement ends once egg and sperm have united. Junior, you are literally on your own from the moment of conception. Through sheer numbers some are likely to survive and reproduce. Some do this once a year, some accomplish this once a lifetime. Other organisms, perhaps the sea turtle for instance, invest a modicum of parental energy into protecting fewer eggs once laid, by digging a nest in the beach and covering it with sand. Once laid and buried, however, junior is on his own to emerge and escape to the sea without being eaten or run over. They produce fewer offspring than the previous example, but still rely on quantity, and leave "quality" to natural selection. Sliding down the spectrum, other organisms, for example birds, have variable degrees of maternal and or paternal involvement in incubating and feeding chicks until they are ready to leave the nest, usually in the order of weeks of involvement with the young. Brood numbers are variable and may be regulated to some degree by amount of food that is available either through maternal energy constraints or through fratricide. Mammals typically invest a significant portion of lifespan and energy towards gestation and childrearing, with litter sizes varying form 1 (cows, human etc) to 30 (mouse). Most litters are limited by the ability to nurse each offspring. There are varying degrees of paternal involvement depending upon the degree of sociality. There are animals that rut or come into estrous or in other words are ready to mate, once a season (wolf, bear), others that have a seasonal estrous cycle with multiple cycles say in the spring or fall(goat, horse), those that rut several times a year (cats, pigs). Most dogs are diestrous in that they have 2 yearly periods of being in heat. Interestingly, rabbits have no estrous periods and are ready and willing to do it like rabbits at any time of the year.

Humans and perhaps great apes have the longest investment in child development. Humans specifically have a very unique estrous cycle. Most mammals have an overt pattern of sexual receptiveness. There are cues given off by the female at certain times that alert the male that it is time to get down to business (or bid'ness). Pheromones, strategic swellings, and unambiguous behavior patterns prevalent in most animals have been replaced in *Homo sapiens* by a covert, mysterious, vague, come hither pattern of recognition of sexual readiness. Furthermore, the frequency of estrous is monthly, but it is not obvious as to when it occurs. What this leads to, some theorize, is a prolonged monogamous relationship or at least serially monogamous relationships. If the male wants to ensure that his genetic material is passed on, he has to repeatedly inseminate the female and fend off all other suitors for long periods of time. To further ensure an elongated commitment, the process in humans is not very effective. Of the hundreds of millions of sperm that begin the journey only a dozen will make the entire journey. If by chance the egg is fertilized, it may not implant in the uterus or may not make it much past implantation due to hormonal or chromosomal abnormalities. In fact, nowadays, a couple is not considered infertile until they have had a year of unprotected intercourse. Thus to ensure paternity, the man literally has to stick it out.

From this pattern of bonding arose the cultural concept of spousal love; a social adaptation to cement the pair bonding and ensure that important survival information from the mother and father was passed on to the next generation. This social survival adaptation of living in groups or packs or tribes, estrously covert with monogamous, yet potentially promiscuous, bonded pairs tightly bound with notions of "love", using tools, and extensively exploiting the environment was and is a vastly successful survival strategy.

Why we have developed this strategy of continuous fertility is a little mysterious. Some in the field suspect it has to do with the fact that our ancestors were an easy target once we decided to exploit the environment away from the safety of the tree tops. Small, slow, olfactorily challenged, bereft of thick hide, claw or tooth, we must have made an easy treat to the carnivores back in the day. In that natural selection crock pot of life on the savanna, we needed to and did develop adaptations to survive and thrive.

Perhaps we had become too successful. Infant mortality based on present day and historical hunter and gatherer societies was likely relatively low. Life expectancy was likely relatively high once predation was minimized via tool use. Disease and trauma probably took their toll but were not usually threats to tribal survival. If there were deaths, it would not take long for the female to be prepared to replace the child at least numerically. These successful adaptations were likely to push the constraints of environmental sustainability. In other words, populations at some point would have placed a strain on the environment to support survival of the group, especially in times of prolonged drought or decreased environmental carrying capacity. There needed to be a governor on population control if, during tough times, the genetic information of the breeding pair or group were to be passed on and not wiped out completely. Other species, like rats, do this by infantile cannibalism or infanticide by dominant males in other species. Because of our development of love and social cohesiveness, this became unfathomable. We are social, continuously fertile, fiercely protective, allegiant and intelligent. Our social, cultural and physical adaptations have not allowed us to easily control our reproduction. Without something checking our numbers, (especially in a pre-industrial, pre-infectious disease ridden, predominantly hunter and gatherer based subsistence pattern) we would quickly over populate. A prolonged drought of 5 to 10 years might completely wipe out a tribe that was too large. The successful proto-*Homo sapiens* had to evolve to not get too big for their collective britches.

It also likely required group cohesion and stability to succeed. Because of the relatively long dependence phase and relative non-productivity of offspring, our species relied heavily upon the wisdom and experience, as well as labor, of parents. Whereas most pack or group animals rely on assimilation of the young to productivity relatively rapidly, our species requires a 15-18 year wait until the young are fully productive and independent. The life expectancy for most mammals is not even this long. Therefore compared to other competing species we live longer, gestate year round, have a year round drive to reproduce, and are relatively successful at surviving to adulthood. From a wonderful compendium of modern day and historic ethnographies of hunter-gatherer societies, Marlowe in 2005 reports that up to 60 % of *H.Sapiens* in foraging societies survive to adulthood.[21] The fertility rate is estimated between five and six children per woman. Even if we assume total monogamy, each couple will raise 3 or 4 children to adulthood which is at least a 50 % but possibly 100% increase in the population each generation. Compare that to chimpanzee mothers on average produce offspring every 4-5 years and on average have 4-5 offspring in a lifetime. It is estimated that 50% of chimps will die before reaching fecundity. Thus chimps have to 2-2.5 offspring survive to fecundity which is barely sustaining the population assuming a relatively equal ratio of males to females. Humans, meanwhile, have the potential to breed like rabbits and exert significant population pressure. Furthermore it is rare for other species to become pregnant while having a dependent infant. In humans this is the norm. Without a mechanism to control population numbers, the fitness of the families that make up populations is at risk.

We have seen that certain adaptations like cancer, genetic disease or altruism might be such a mechanism. This is crucially important for *Homo sapiens* who have other highly-invested offspring who may be weaned but are far from being independent. Infant mortality (or grandparent mortality) might not be such a bad thing in terms of focusing energy availability on the established young children during times of resource constraint.

There are other quirks of the human body that are perplexing. It seems that our own physiology is out to get us. Or at least that we are innervated with the blue print for sacrifice in our very DNA. If it seems that this is an extraordinary claim, (perhaps to the strict Darwinist), then perhaps I can offer some extraordinary, if admittedly circumstantial, proof. Perhaps this is the bias of a physician who constantly sees the human body at its worst; or perhaps there is a reason, under natural selection, why we still have appendicitis, strangulated hernias, and a gallbladder that can kill in 3 or more different ways.

PART II
The Sacrifice of the Gene

Chapter 1

APOPTOSIS AND TELOMERES
The blueprint for sacrifice

Once life evolved to multicellular somatic structures and cellular moieties with specialized functions, there needed to be command and control mechanisms to insure stability of forms. With cellular specialization, eventually organisms evolved to the point where no individual cell could survive on its own. For instance, locomotive cells could not be expected to be efficient at fending off potential pathogens. Digestive cells could not be expected to be efficient at locomotion. Cells that were exposed to the slings and arrows of extrinsic attack had to be able to renew themselves lest their function would be short lived. For cellular differentiation to occur, a command and control architecture had to evolve. For those tissues whose cells required turnover for efficient functioning, there needed to be a way to control the turnover. If the turnover process were too exuberant, the organism risked channeling resources inefficiently. In other words, specialized cells had to be capable of dividing but not overriding their function. Immune cells for example, have to be able to turn on their function, rapidly divide, and once the task is completed, be able to turn off again and not seriously harm the organism itself. For embryonic development to occur there needed to be a way to mold the protoplasmic blob of the fetus into limbs, fingers, toes, hooves, or wings.

Injured, non-functional tissue also has to be dealt with as well. When an injury occurs, the body removes the tissue and cellular debris in a much different way than it removes tissue from a developing embryo. With injury, the smallest blood vessels are disrupted leading to leakage of warm blood into the tissue. Without the delicate, intact system of capillaries, tissue hypoxia ensues around the injury. Without oxygen, cells swell and burst releasing strong signals for immune cells to swoop in and clean up. Immune cells release powerful peroxidases and other toxic chemicals to clean up the injured and potentially germ-filled region. This leads the four horsemen of inflammation, tumor (swelling), calor (heat), rubror (redness) and dolor (pain) to ride in. The cells themselves, if examined under the microscope, would look disheveled, swollen, fragmented, and generally not themselves. The word for this type of cell death is necrosis, from the Greek word for death.

This type of reaction makes sense. The redness and swelling is from an increase in blood flow and a general leakiness of the surrounding capillaries so that more immune cells and proteins can enter the arena to help get the healing process underway. The small blood vessels of the skin usually keep excess heat from escaping our core by shunting blood away from our outermost layers. Inflamed tissues bypass this thermoregulatory mechanism and allow our inner heat to bathe the injured area so that the healing milieu is at an optimal metabolic temperature. The small sensory neurons in the area are affected by the by-products of the inflammation and signal the brain to lie still to avoid exacerbating the already injured tissue further and allow broken bones or soft tissue to heal.

As effective as this process is at healing a cut or broken bone or dealing with a bad infection, it is way over the top in regards to cleaning up the injured tissue. It's like hitting a fly with a sledge hammer. It may achieve the desired effect but at a cost to the surrounding tissue. In fact, the body's reaction to certain bacteria can lead to a kind of China Syndrome, where death of a host ensues not from overwhelming bacterial infection, but from the consequences of the body's response to the bacteria. The immune cells secrete so many of these inflammation signals that the lungs, kidneys, and central nervous system think they are under attack as well and shut down leading to hypoxia and death.

The organism needs a more subtle mechanism to affect change. With inflammation and cell necrosis, the cell membranes of injured cells are torn asunder. Potent pro-inflammatory molecules are thrown to the wind and a defilade of wanton, toxic, immune mercenaries descend on the battlefield killing friend and foe alike. When the body needs to quietly turn over unneeded tissue, it relies on cell suicide.

The term apoptosis (Greek for falling or dropping off, like leaves or petals) was originally described in 1842, but because the process happens so quickly and quietly, it was not until the electron microscope was widely employed in the 1960's that scientists could accurately corroborate the work of the earlier scientists.

The process relies on the fact that every cell has a built in suicide capsule that it can take when necessary. The vital component of almost every cell is the mitochondria, the powerhouse of the cell, where energy is created. If this organelle is targeted, no energy is available and the whole cell will shut down. In the event of a stimulus, the cell signals the mitochondria to shut down. But instead of exploding and releasing its cellular shrapnel to the wind, the cell quickly and quietly involutes into small quanta of broken down proteins and nucleic acids surrounded by pinched off bubbles of cell membrane. These vacuoles are then engulfed by immune cells and the contents recycled without inciting a whole lot of fuss.

Every cell contains this suicide pathway which can be initiated by a host of intrinsic or extrinsic factors. During development, the interdigital cells of the human embryo undergo apoptosis under the direction of certain hormones, to allow the formation of distinct, independent fingers and toes. When it doesn't happen properly the webbing remains at birth. Radiation, hormones, drugs, excessive heat, or damage to the genetic material all can initiate apoptosis. The process begins with a stimulus. In the heart for example, a heart attack or myocardial ("myo" muscle "cardia" heart) infarction (death) occurs when the blood flow to the heart muscle is shut off rapidly, usually as the result of a rupture involving the wall of an abnormal cardiac blood vessel. In response to the lack of blood flow and oxygen, the cells suicide genes are activated. It seems to be saying, "Times are going to get tough. Before I use up precious energy trying futilely to stay alive, I will sacrifice myself in the hope that others may survive"

It's the body's way of making sure that old worn out parts are turned over without making a big fuss. It's a way of ridding the body of potentially dangerous cells. Lest a cytotoxic white blood cell cross the Rubicon and attack other healthy cells, they have installed in their genetic make-up a self destruct mechanism that keeps their numbers in check. We don't need you anymore; we need your proteins and carbohydrates somewhere else. If the process doesn't occur in the lung and elsewhere, tumors can develop from cells that are programmed to divide, but have lost the signal to reign in their cell division.

For some cells, this process is activated by extrinsic causes. For some cells, whose death is required for efficient organ function, the process can happen automatically. How this works is a very fascinating story. The full details are beyond the scope of this work, but a brief foray into the story might be illuminating. The story begins close to the beginning of the evolution of complex organisms.

Bacterial DNA is circular; the protein machinery for replicating bacterial DNA has no problem faithfully accomplishing the replication process because there is no beginning and no end of the track (the DNA strand) on which the proteins work. The train-like replicating protein can continue on around the track until the task of creating a new strand is complete. Organisms with nuclei, eukaryotes, have by and large evolved linear DNA into one or more lengthy strands called chromosomes. The problem with linear DNA is that it cannot be faithfully replicated time and time again because the protein machinery would run out of track before the last few nucleotides could be replicated. Over time and many cell divisions, the DNA would be degraded and important information would be deleted. To overcome this problem, organisms have evolved a dunce cap for the end of each chromosome. Called a telomere, the dunce cap is a repeating pattern of DNA segments at the end of each chromosome that don't encode for any proteins and are in place to protect the important encoding DNA from eventual degradation. Like a knot at the end of a rope, the telomere protects the encoding DNA from fraying or end degradation. But it doesn't last forever. Once the telomere has been sufficiently degraded, the chromosome cannot replicate further. The cell is recognized as damaged goods and is subject to apoptosis.

Thus the consequence of evolution from a circular chromosome in the bacteria to a linear chromosome in eukaryotes was genetically programmed death. It becomes a "chicken and the egg" quandary. Did we evolve more complexity which required linear chromosomes to allow command and control, or did we evolve linear chromosomes as bacteria which then allowed complexity to ensue? There is no doubt, however, that the command and control necessary for integration of function of disparate cellular entities required the evolution of cell death for efficient function.

Thus, down to our very marrow, we are programmed for sacrifice. Human physiology employs cells that expire once their job has been performed. It is efficient and necessary for the good of the organism that this should occur. Extending this metaphor to macro-evolution and the development of senescence and programmed organism death might be appealing for the group or multi-level selectionist. The cell (individual) subjugates its survival or fitness for the good of the organism (group). However, this is a bridge too far. It does not imply that we as humans are analogous to the cells of some "superorganism" or supernetwork; it simply means that the genes that are driving our cellular physiology have the same game plan as the genes that determine our longevity. The suicidal *modus operandi* that is in place at the cellular level can easily be tweaked to control an organism's longevity, should the environment require earlier death for efficient gene propagation. Thus, the notion of sacrifice is well entrenched.

Chapter 2

SYMBIOSIS AND EVOLUTION PROCEEDING FROM COLONIES

From studies performed on mitochondrial DNA harvested from humans from all corners of the planet, scientists have discovered that we, as a globally diverse species, all descended from a single great, great,…. great grand mere only 200,000 to 160,000 years ago. That is not very long ago, relatively speaking. Hominids have been roaming the earth for over 5 million years.

There is debate over the methodology of the experiment and the interpretation of the data, but subsequent experiments on genes of the Y chromosome and different aspects of the mitochondrial DNA have either corroborated this early date or even proposed an earlier date for our most recent common male ancestor. Physical anthropologists postulate that we, as a species, emerged from Africa as recently as 60,000 years ago.

Theories and corollaries of evolutionary thought seem to assume that natural resources are abundant. It is a consequence of thinking that "all things being equal, the fastest animal will survive" or the strongest or whatever trait which is being selected for. In other words, in a resource neutral environment, the best climber will get the most fruit or the strongest, fiercest predator will get the most game and be able to reproduce the most often.

However, environmental pressure is never equal or neutral. Resources are characterized by scarcity punctuated by temporary abundance. Think fruiting trees, plankton blooms, calving seasons, egg-laying seasons, etc. Therefore, it is not always the strongest or fastest animals that survive, but those most able to cope with prolonged dearth and those most able to utilize punctuated, ephemeral abundance.

Add to this, the chaotic occurrence of prolonged drought, over many years perhaps, where the usual abundance is absent for a very long period of time, and you have a scenario of wide spread elimination of vast numbers of a species, if not downright extinction of all but the leanest members. In this setting it is perhaps not the strongest who survive but the quirkiest, the most fortunate who survive.

Perhaps this quirkiness of nature overlies the origin of symbiosis as well. Take the termite. The termite relies on a gastrointestinal protist (kind of like a bacterium) to breakdown the cellulose or the woody starch in your brand new deck. Presumably the termite did not seek it out to be able to eat wood. Likely the termite forefather ate something else that it could digest, but one day got colonized with this protozoa. So long as the termite's original food was available, say nectar, this protist was a hitchhiker, or a parasite, robbing the termite of some of its food, but not outright killing the animal. Now it happens that there is a blight or drought so that few resources are available for years. This termite then, thanks to its capable intestinal hitchhiker, has the ability to survive on wood which has not been affected by the blight or the drought. He finds he can feed others by regurgitation or defecation which results in transmission of more protists to others as well as providing nutrients to the colony. This colony survives the blight and as they say in the movies, the beginning of a beautiful friendship.

Perhaps this is overly simplistic, but it remains a plausible mechanism for the development of traits that aid survival that in and of itself is not a positive survival trait. When the termite ate nectar, the protist was survival neutral, if not survival negative, if it ate some of the sugars that would have gone to the termite. However, it became a positive advantage to the termite during times of stress. The termite species of today evolved from this one colony with its fortuitous symbiote.

What does the termite have to do with human evolution, you ask? Well, if the scenario for termites seems plausible, that a major event like a prolonged drought or blight led to evolution of a species from a single colony of termites, then perhaps human evolution has sallied forth from a similar mechanism. We have evolved from a colony of ancestral *Homo sapiens* that were adept at surviving when the environment was at its least fruitful. The fact that our most recent common ancestor is relatively young speaks toward the evolution of a species on the brink of extinction during cycles of relative abundance followed by prolonged environmental dearth. Are there traits in humans that on the surface appear survival neutral or negative like the enteric protists of the ancestral termite, but have survival advantages during times of scarcity?

Chapter 3

THE APPENDIX

The digestive tract is a wonder of coordination, efficiency and function. Digestion begins in humans as soon as we put food into our mouth. Wonderfully efficient teeth grind even the toughest foodstuffs to initiate the breakdown of carbohydrates, protein and fat. Breakdown of starches even begins in the saliva. So called amylases cleave the long branching starch molecules into cellularly more manageable individual sugar molecules. By a complex neurologic process that most people never give a second thought to, we chew our food into little packets or quanta or boluses which pass dangerously close to our windpipe and into the esophagus fed by a conveyor comprised of the tongue and other wonderfully coordinated jaw and neck muscles. So efficient is this mechanism that we can do it standing on our heads with the thinnest of liquids and not spill a drop into our nose, mouth or windpipe.

Once into the esophagus (essentially a muscular tube lined with cells remarkably similar to skin cells) the food is propelled into the stomach in prepackaged units ready for a second phase of digestion. In order to winnow the wheat (food) from the chaff (bacteria, germs) the stomach bathes the food in the strongest acid in nature, muriatic or hydrochloric acid (pH 1.0). This does two things; it kills most of the bacteria that swarms the food we eat and it initiates breakdown of proteins into their component amino acids so that at the cellular level, that great big wad of strip steak you just inhaled can be absorbed at the microscopic level and be packaged up to be sent out to where it is needed.

Industrial strength acid in everyone's stomach? No wonder we get tummy aches from time to time. Talk about your double edged sword. The stomach lining though has a good defense. Certain cells secrete a carbohydrate rich slime that coats the stomach and is impervious to the acid. Aspirin and other similar compounds and certain acid- loving bacteria are thought to impede this barrier and lead to ulcer formation.

The stomach's other function is to churn and mix our food with the acid and certain stomach enzymes as a prelude to absorption of the nutrients which takes place in the small intestine. There is a muscle at the end of the stomach that stays shut most of the time allowing the churning process to continue and ever so often opens slightly to allow slurry of stomach succus to pass through. Oddly enough, no real absorption of food takes place in the stomach, with the exception of alcohol, which can pass through or diffuse through the slimy coating of the stomach and be absorbed directly into the bloodstream. This is why if you drink on an empty stomach the alcohol can get into your system quicker, because there is no food to get in the way of its absorption.

Another function of the stomach is to be a compliant pouch to allow you to stuff your face and have a place to hold it. Your stomach can expand to three or four times its resting size. And much to the chagrin of dieters everywhere today, the signal to let you know you couldn't possibly eat another bite doesn't get to the conscious brain for twenty minutes, allowing you to pack in that extra cheeseburger after the stomach has already cried uncle. I suppose it is evolutionary advantageous to be able to absolutely stuff yourself like a Viking on holiday.

After the food is churned and spit out into the small intestine, a host of enzymes, antacids and emulsifiers further breaks down the food we've eaten into its basic components which can be absorbed by the billions of cells that line the approximate 25 feet of small intestine. Proteins and sugars are sent directly to the liver to be further processed, detoxified, and stored or shipped out to the rest of the body as needed. Most fats, interestingly enough, escape initial processing in the liver and are packaged up and diffuse out to the entire body after a meal. This is why if your doctor orders a cholesterol test, he wants it to be a fasting blood draw, so that the cholesterol that we have just eaten won't show up in the blood that was just drawn and falsely elevate the results of the test.

The end of the small intestine is the beginning of the large intestine or colon. The last part of the small intestine is the grand recycler of the body. It takes about 6 liters of fluid a day to emulsify and digest our food and neutralize the acid from the stomach. Most of that fluid and the salts needed to emulsify ingested fats are reclaimed by the last few feet of intestine. Were it not for this ability of the small intestine to recycle, we would all have a massive case of the flux. And we'd be pretty thirsty all the time too, or dying of dehydration. Thank you small intestine.

The job of the colon is to wring every last drop of water from the waste products of digestion and to bundle it up into discrete or not so discrete packages for elimination. One of the other jobs of the colon is to harbor billions and billions of bacteria (Do I sound like Carl Sagan?) which believe it or not is good for us. These good bacteria living in the colon don't allow the evil nefarious bacteria to gain a foothold in this nutrient rich environment and invade our inner systems. They also break down certain byproducts of digestion and provide a wonderful odor to emanate from our backsides. The trade off by having all these little stinkers present in our colon is to be able to digest things we ourselves wouldn't be able to digest. The end result of all these critters is death and elimination. Even in dying however do they provide a useful function. It's thought that up to half of the bulk of stool is comprised of defunct bacteria and in passing they allow the colon to be swept clean of some of the toxins associated with digestion. Let's face it we put a lot of yucky stuff into our bodies. If bad stuff hangs around too long (think brother-in-law) it could cause a lot of problems.

Interestingly, the small intestine doesn't empty directly into the beginning of the colon. The colon itself is like an inverted "U". The first part is called the ascending colon, well, because it ascends up the right side of the abdomen. The colon then takes a 90 degree turn at the liver and doglegs more or less across the top of the abdomen until hits the spleen and then sharply descends down the left of the body (that's right you guessed it-descending colon) until it hits the "S" shaped portion called the "S" colon. No just kidding, it's really called the "sigmoid" colon from the propensity of early anatomists to use the much cooler Greek alphabet when naming bodily objects. Then things get interesting in the rectum or as I like to call it, the colonic lobby. Here stool hangs out until a person sees fit to "drop the kids off at the pool" (or insert other salacious scatological euphemism here.) (Note to self: Get rid of that Word of the Day Toilet Paper) The small intestine empties into the large intestine about ¼ of the way up the ascending colon. This creates a blind cul-de-sac, if you will, (or dead-end, if you won't) called the cecum. That brings me to that goofy little excrescence dangling off the end of the beginning of the colon, the appendix.

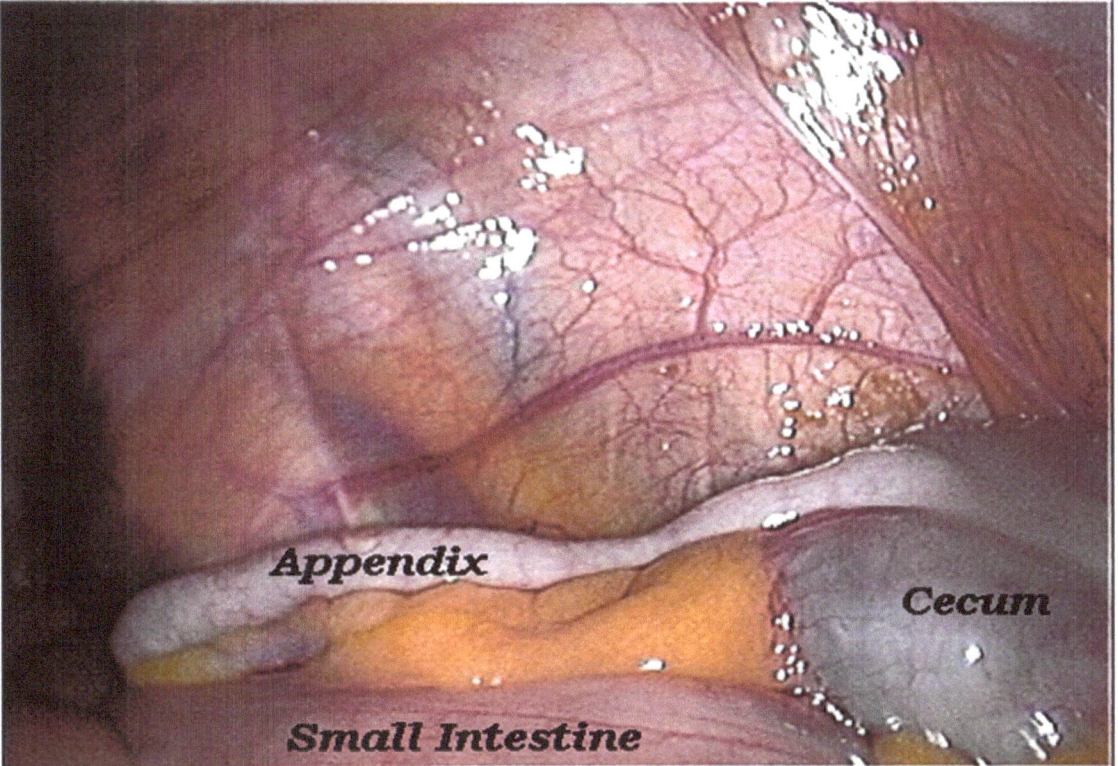

Figure 8. Laparoscopic view of a normal appearing appendix.

From the Latin, *appendare*, to add on to, it, like its etymological namesake, is an afterthought, a vestigial hanger on which has no rigorously discernable function except perhaps to kill us. It does look like a worm, hence the medical name, *vermiform appendix*, or worm-like appendage, from the outside, but in reality is a blind ending tube that has a narrow opening into the cecum. (General Surgeons have argued its existence is to keep their kids in Ivy League Schools, but with medical reimbursements being what they are today, and its propensity to go bad in the very middle of the night, it really is a nuisance organ to physician and patient alike.)

If you look at it under a microscope it has a lot of white blood cells hanging about just underneath its lining, which, especially in the childhood form of appendicitis, can swell up in response to infection. Too much swelling and the lumen, the inside of the tube, gets blocked off. The appendix can't empty and the bacteria in the appendix behind the obstruction can party like teenagers on spring break. This leads to swelling, distension and can lead to eventual rupture or bursting of the appendix like a ripe melon left out in the sun. Spillage of fecal matter into the abdominal cavity causes some of the most intense pain discernable by humans. It also leads to blood poisoning, organ shutdown, and death if not treated.

Researchers have sought for explanations of the persistence of this wormy tissue, but none seem to pass the sniff test. Most scientists believe that the appendix is a vestigial remnant of a once proud cecum that harbored bacteria responsible for breaking down plant material into absorbable nutrients. In some animal herbivorous species the appendix is very long, large and harbors bacteria with the enzyme cellulase, which allows them to convert the plant cell wall (cellulose) into glucose, which it can use. No mammalian cell has the enzyme to do this. Even the termite can't do it alone; it requires a humble protozoan to break down your wood deck. (Or perhaps the protozoan needs the termite)

There may be some immunologic function that the appendix that has not been discerned. Beneath the lining of the appendix there is a host of white blood cells that could function as a weigh station for the immune system that keeps the mercenary bacteria of the colon in check.

Some have postulated that it persists because it performs a kind of a bacteriological reset button; if the colon gets overwhelmed by a bad infection, the body is able to fight it off, but if in the process most of the good bacteria are killed off, the appendix, which is in the cul-de-sac off the fecal highway, can repopulate the colon with its normal bacteria. But the fact is you do just fine without one. Most carnivorous animals don't even have one. Dogs don't, cats don't, so why do we still have one?

It is estimated that the lifetime risk of developing appendicitis is 7 % and that before surgical therapy was perfected, acute appendicitis was usually fatal. What is the advantage of the appendix? Are we on our way to getting rid of it, evolutionarily speaking?

Let's look at human evolution in general to answer that question. Proto-hominids diverged from great apes approximately 6-7 million years ago. Since modern day apes still retain an appendix, it is safe to assume our proto-hominid ancestors possessed one as well. Assuming the appendix conveys no real survival advantage to the individual (because removing the appendix has no decrement on life expectancy) and without modern medicine, would kill, conservatively, 2-3% of healthy individuals before reaching sexual maturity, one would postulate that over time, this little ticking time bomb would have been eliminated from our genetic blueprint.

But it hasn't. Indeed evolutionary pressure to keep the appendix seems very strong. For an organ that seems an afterthought and seemingly has no real purpose, it is amazingly persistent. It is extremely rare to be born with a normal intestinal tract *sans appendix*. It is estimated in the surgical literature to be absent from birth in about 1 out of 100,000 people.

One could argue that it is an evolutionarily neutral adaptation. There is not significant penetrance of appendicitis to weed out the appendix from a genetic standpoint. In other words, appendicitis is not a foregone conclusion if a person inherits an appendix. Indeed, most individuals with an appendix do not develop the life-threatening disease. Therefore there is not enough survival pressure to eliminate it. There are too many people reproducing happily with appendices to get rid of it.

One could also argue that the current prevalence of appendicitis could be a relatively new phenomenon. That is we only know the rates of appendicitis for the last 200 hundred years or so. Our pre-historic diets could have resulted in fewer instances of the disease. Indeed there is ethnographic evidence of hunter and gatherer societies with diets high in fiber and low in fat that shows that appendicitis is relatively rare.

But it is not non-existent. Even apes get appendicitis. Alterations in diet might explain why adults get appendicitis; what with our low-fiber, white bread, beady stools clogging up the appendix. A child with appendicitis, though, doesn't typically have a fecalith or stool ball clogging up the appendiceal opening, but more likely has the white blood cells underneath the lining swelling and obstructing the lumen. Without modern medicine or autopsy series it is difficult to put a number on the incidence of childhood appendicitis in historic hunter and gatherer societies.

I would argue that it remains as an adaptation to prevent overpopulation in a scarce resource environment. The evolutionary pressure to rid the GI tract of the appendix is outdone by the evolutionary pressure to keep it. Remember that agriculture and domestic animal husbandry have only been in place for the last 10,000 or so years, which is a wink in time of our evolutionary development. Seven percent is not a big number; appendicitis is not going to wipe out the whole clan, but it will cull the herd, so to speak, allowing the rest to be well nourished and have strong offspring. Less is indeed more.

Though appendicitis does not appear as a recessive trait, the prevalence of death from appendicitis is similar to the penetrance of a "malignant" recessive trait. In other words, the gene for the persistence of the appendix can exploit the fact that a small number of people will die from appendicitis. In populations with persistent appendices, more people will survive than in populations without appendices, because those with appendices keep their numbers at an efficient level for gene survival.

Imagine two families with 10 children. The Conapps all have their appendices. The Sinaps have lost theirs. Two of the Conapps' children have died from appendicitis, leaving them with eight. None of the Sinaps die from appendicitis. They then both face a long winter. The remaining eight Conapps all survive with what stores they have to just barely make it through. In trying to keep 10 children alive all winter, all the Sinaps are weakened by the protein deficit and half are killed in the last two weeks of winter and the remainder are too weak to adequately forage.

Perhaps this stretches the imagination a bit too far. The above example is exaggerated to make a point. Over several lifetimes, however, the death of a few could translate to the survival of many.

The appendix's persistence results in random reduction in the number of a species even when times are good and food is plenty. Thus we don't have to wait until pressure for scarce resources becomes critical before the individuals of a species start dying off. By the time resources get scarce, the whole clan is weak and susceptible to being wiped out as a whole. But by always culling the herd, the herd stays strong. As an evolutionary pop-off valve, it keeps environmental pressure in check.

As another metaphor, imagine a scenario at sea, where a group of cutthroat pirates needs to get to the next port of call where a large treasure awaits. They are uncertain of the distance, the time involved, and whether they will be able to provision along the way. They also realize that other pirates are after the same treasure. They have just taken a prize merchant ship and are faced with a conundrum. Should they take on extra crew members to sail quickly to the next port of call? In doing so, not only do they risk running out of provisions with such a large crew meaning no pirate would make it to the treasure, but they also risk having to share the treasure with a multitude of other pirates. Or should they keep the crew small and risk sailing slowly but remain unlikely to run out of provisions? Here they are weak from attack from other pirates or other dangers at sea. Through previous similar trials, the pirates have evolved an ingenious plan; they take on a large crew and sail quickly at first. As provisions start to peter out and can't be refreshed, they start to throw overboard the oldest pirates who aren't pulling their weight and the rawest land-lubbers who aren't worth the grog. To further ensure that this ship doesn't run out of provisions, the pirates have developed another callous but practical implement. The crew holds a lottery which results in the random shooting of one pirate per week. This ship with this strategy is able to beat out all the other pirate ships with less cutthroat policies. (Incidentally, ritual acts and superstition would accompany the performance of the lottery to try to influence the process of random crew reduction.)

The ship that wins the race is the one that has a propensity to press as many sailors as would seem practical (or stepping out of the metaphor, the breeding pair that optimizes offspring survival has a strong desire to reproduce), has the ability to winnow the crew (has intrinsic causes of death like cancer or genetic disease) and can randomly reduce the population of even seasoned sailors (possesses attributes like an appendix that keep the brood size and the population size at a reasonable level).

The metaphor of pirates at sea may seem like an analogy for multi-level selection. Pirate ships that efficiently band together out compete those others ships who are less efficient. But the metaphor of the pirate ship refers to an individual (or at most to a breeding pair). The individual crew members perhaps represent an animal's offspring. The daily orders for running the ship and the orders for reducing the crew represent an individual's genetic messaging. The treasure for us is the access to food that, like the treasure for the seagoing pirates, is neither abundant nor readily available, but leads to the ultimate prize of survival of offspring. For pirates and individuals on Darwin's stage, both must survive long distances and seeming epochs of wanting before they reach the fruits of their labor.

It's a cruel but effective ship, and the methods are passed on to the young pirates who go on to lead their own ship. Nature, unlike the pirates, is not subject to our codified mores or articles of piracy. Rather than cruel, nature, in the words of Richard Dawkins, is "pitilessly indifferent." As tough as such rawhide may be to some to digest, DNA is as sympathetic as a spreadsheet, yet wonderfully adept at molding the entire planet.

What other part of our anatomy is aiming at us?

Chapter 4

TESTES ON THE OUTSIDE

Why did the male of our species evolve with the most important (evolutionarily important) and most sensitive equipment dangling on the outside of our bodies without protection and sinisterly vulnerable to the knees and implements of social and athletic attack? Surely it is a cruel design flaw, to have the testicles on the outside. The Indifferent Evolver must have been a fan of slapstick video programs.

Let's look at the whole process of just how this happens and why it might be advantageous.

The testes in human beings and all mammals start on the inside of the abdominal cavity near the kidney in the embryo. In some mammals, they stay on the inside. Animals like elephants, rhinos, shrews and marine mammals keep their man parts on the inside. However, most have adapted mechanisms to keep the sperm producing cells of the testes cool, or at least cooler than the rest of the body. Some mammals do this by cooling the blood that flows into the testes by placing these blood vessels next to the skin Other mammals keep'em cool by positioning them outside the steamy abdominal cavity, and regulate the temperature by regulating the distance from the body's heat source. Closer when colder, further away when warmer.

But how do they get outside the body? It seems that under the influence of testosterone the developing gonads in the embryo migrate downward toward the groin pulled by a tether-like structure called the gubernaculum and pushed perhaps by abdominal pressure. This push-pull process eventually results in migration outside the body through the muscle and connective tissue of the abdominal wall and into the scrotum. Once migrated outside the abdomen, the lining of the abdominal cavity, the peritoneum, pinches off and closes the defect so that the floppy organs (intestine) and fat inside our bellies cannot hitch a ride. Still tethered by some of the muscle that it picks up along the way, the testicle can move up and down within the scrotum.

Why go through all this trouble of pushing them to the outside? Well, intuitively from the male perspective, the more sperm you make the more likely you are to pass on your genetic material. Thus evolutionary pressure favored the individual whose testicle could turn out the most sperm. It turns out that it is more efficient if this process occurs at a lower temperature. It may lead to fewer mutations of the vital components of the spermatozoa and thus be advantageous. Such a metabolically active process generates a lot of heat. Some species dealt with this by internally cooling the testes with cooled blood that passes near the skin before reaching the gonad, similar to a car's radiator. In other species, it became more efficient to move the process off site to a cooler locale with a fairly simple thermoregulatory system. In some, (hedge hog) this is the inguinal canal (the groin crease), in others, the scrotum, with its plethora of sweat glands and paucity of insulating fat, became a perfect testicular oasis.

In the female of the species, this is not an issue. The ovary is not nearly as metabolically active. Quiescent in most mammalian species for most of the year, the female gonad was perfectly content to remain in the abdomen. (Not really a volitional choice, per se, but there was no evolutionary pressure for the ovary to migrate.)

For most quadrupeds, this was a win-win situation. The testes were happy on the outside, and the path of emigration was in a relatively innocuous location at the apex of the abdominal cavity where the leg meets the abdomen at a distance from where the abdominal pressure is greatest near the mid abdomen. The inguinal canal or the pathway of testicular migration (and subsequent potential weak spot) is directed uphill in the quadruped. Inguinal hernias do occur in pigs (in about 1-5 %) and horses but are rare in cattle. In no species is it as common in humans (about 25 % lifetime risk for males).

For bipedal mammals, namely us, punching a hole in the lower abdomen so our man parts can chill becomes a big problem. Our upright stance combined with gravitational vectors place the inguinal canal right in the crosshairs when we lift, strain, cough, poop, pee (can you tell I have two little kids) or sneeze. First of all, thanks to our quadripedal ancestors who had no need for a strong lower abdominal wall, the lower abdomen in mammals doesn't have the connective tissue strength the upper abdomen has. Secondly, the pinching off process of the abdominal lining, described earlier, doesn't always occur correctly. When it doesn't, babies are born with a congenital hernia. When the naturally weak spot gradually wears down over time from incessant pressure of everyday bipedal living, it is an acquired hernia. In both instances, loops of intestine can get in the hernia and fill it up like clowns in a Volkswagen. Too much gets in and it can't get back out. The trapped, or incarcerated, intestine swells and swells and the pressure builds such that the blood supply to the intestine is cut off. This is a strangulated hernia and without prompt surgical therapy it is universally fatal. It is estimated that 1-5% of unrepaired hernias will go on to strangulate.

Again why do we persist in having this unusual testicular arrangement that can potentially kill us? Why is there no evolutionary pressure to strengthen the lower abdominal wall or to keep our daddy parts on the inside? We (our ancestors) have been upright for at least 5 million years. Why do we still get hernias?

It could be that this is just how things are for humans. Evolution doesn't always make sense. Ancestrally, we developed efficient, if wandering, testes before we developed bipedalism. With the outside teste already in place, the move to upright posture was not a significant enough stimulus, evolutionarily speaking, to override the testicular odyssey. In other words, if it isn't significantly broke, don't bother fixing it. In fact, if a testicle doesn't descend into the scrotum, it has a very high rate of diminished or absent sperm production. So any mutation towards nondescending testicles is likely not to be passed on.

However, here are the numbers on children with congenital hernia. Up to five percent of the population is born with one (or two) and up to 30 % get incarcerated which was a death sentence before modern surgical techniques. That's 1.5 % of folks who will not pass on their genetic material. While 1.5% may not sound like a large number, over time, given thousands and thousands of generations, mathematically speaking, congenital hernias should have been eliminated a long time ago. And even if it doesn't incarcerate or strangulate, the hernia will enlarge. I can't imagine it's too easy to survive and reproduce in pre-historic society with a giant inguinal hernia.

Furthermore there are biochemical explanations for why some people are prone to forming hernias. The strength of connective tissue and scar tissue is dependent upon the body's ability to interweave or crosslink collagen. Collagen is found throughout the body, in several forms, but is the main scaffolding protein of our ligaments and tendons. Patients with genetically impaired collagen synthesis are much more likely to develop acquired hernias. Patients who cannot crosslink collagen normally have a higher incidence of hernia and hernia recurrence once surgically repaired. Thus some folks are genetically primed to have weak connective tissue and thus to form hernias.

It doesn't make sense unless there is some advantage to pass on the propensity to develop hernias. It seems plausible that there is a survival advantage for individuals if we don't get too big for our collective britches. Weak abdominal walls and circuitous testicular wandering with potentially disastrous consequences persist in us so that we do not overwhelm our species' capacity to survive within the resource restraints put upon us during the evolutionary development of human beings. It remains as a pop-off valve for our incessant drive to procreate, lest we become too successful at fruitful multiplying.

Evolutionary pressure to strengthen the lower abdominal wall is not enough to surpass the pressure to keep it the way it is. The gene for weak abdominal walls and hernia formation persists to keep offspring numbers at the optimal number for breeding pairs and the population in general.

Chapter 5

THE PIMA INDIANS AND GALLSTONES

The gallbladder is a physiologic relic of a time when food was scarce and spoiled quickly. This meant that once upon a time our ancestors experienced long periods of relative fasting punctuated by periods of Bacchanal, rib splitting, belt-loosening gorging to be able to survive. A ready supply of digestive bile came in handy for these type situations. The gall bladder does nothing more than store and concentrate bile made by the liver and empty it into the digestive tract when presented with a large fatty meal. (Bile is an emulsifier, allowing the fats we eat to mix with our bodies which are 75% water) The problem with this arrangement is that nobody eats this way anymore (with the exception of a few sacred holidays, and the Superbowl). We eat three to four small meals a day and the gallbladder rarely gets a good workout any more. It becomes a flabby, weak, girly-man appendage and doesn't empty properly.

So what, so my gallbladder doesn't empty like a Viking's, you say, what's the big deal? Well it can be a big deal. Bile is composed of bile salts, like soapy water is composed of soap salts. If the bile becomes stagnant then those salts begin to precipitate out or congeal into tiny or sometimes not so tiny stones. Most of the time, these stones are innocuous. In fact it is estimated that 10 % of Westerners have gall stones and don't even know it. However they can really cause problems if those stones get out of the gallbladder itself and get stuck.

The gallbladder empties into a small soda straw like opening called the cystic duct which in turn empties into the common bile duct. The CBD brings bile from the liver and gallbladder into the intestine. Right before the CBD empties into the intestine it is joined by the main duct carrying digestive proteins and digestive juice from the pancreas. (See Figure 5)

If the stone gets stuck in the cystic duct, the one leading out of the gallbladder, the gallbladder blows up, the bile behind the stone in the gallbladder can get infected from bacteria from the digestive tract and a serious infection can set in. If unrelieved, the blood supply to the tense and swollen gall-bladder is cut off and the wall of the gallbladder wall dies and ruptures. This spills infected, caustic bile all over the peritoneal cavity and can result in peritonitis and death.

If the stone intermittently gets stuck and pops back into the gallbladder when the pressure increases, the patient feels pain under the right ribcage after eating and describes this as a gall bladder attack. Physicians describe this as biliary colic or symptomatic cholelithiasis.

If the stone gets stuck in the CBD, bile can't get out of the liver which causes the patient to turn yellow or in medical parlance becomes jaundiced. The bile can get infected behind the blockage resulting in a severe liver infection called cholangitis. This causes a high fever, pain, and a detrimental chain reaction all over the body (systemic) to the bacteria that get in the bloodstream that is, without antibiotics and ICU type care, universally fatal (called sepsis or sometimes referred to as blood poisoning).

If the stone gets lodged where the pancreas empties into the CBD, the pancreatic juices back up into the gland itself and into the blood stream where the proteins or digestive enzymes meant to digest the food proteins we eat, get confused and start digesting the pancreas itself. It's like a small scale digestive China Syndrome, where this leak of enzymes causes a bigger leak which causes more self digestion which causes a bigger leak of enzymes and so on and so on until a chain reaction occurs that causes significant tissue injury and can lead to death if the body cannot contain the pancreatic reactor leak. This is called gallstone pancreatitis and if certain criteria of illness are met, is 80-100% fatal even in the best ICU.

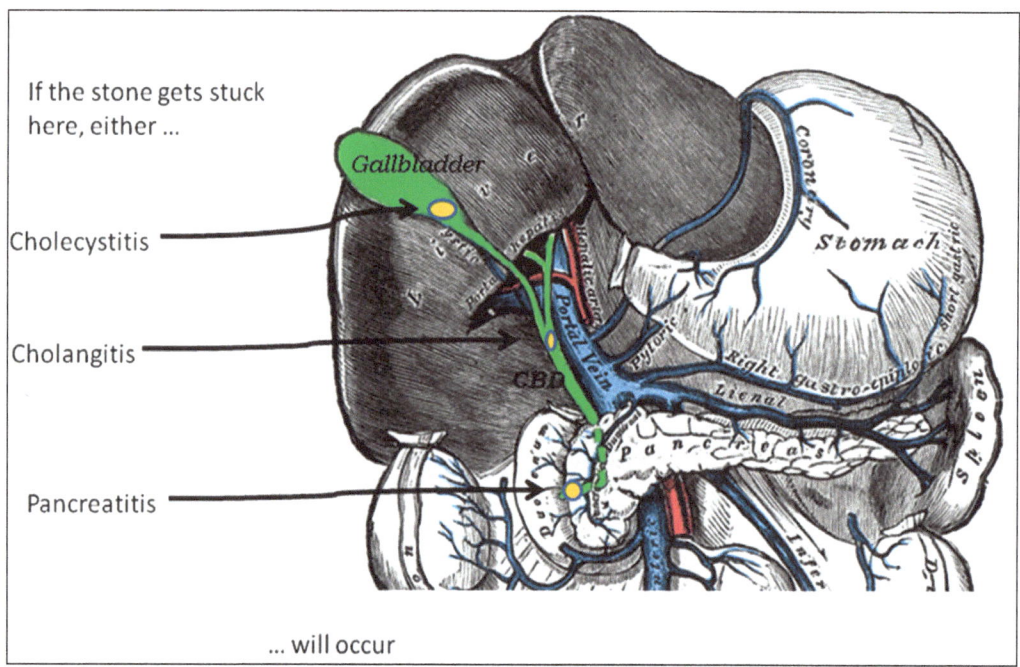

Figure 9. Depiction of gallbladder and biliary anatomy. Modified from the original plate from Gray's Anatomy. From the online edition of the 20th U.S. edition of Gray's Anatomy of the Human Body, originally published in 1918. The liver is seen lifted upwards and the duodenum, stomach and colon have been divided for clarity.

Fortunately most stones either stay in the gallbladder or pass harmlessly and the risk of developing these complications is fairly low. However, if you are a Native American in the desert southwest your chance of developing gallstones by age 30 is much higher. Almost 80% of Pima women develop gallstones and frequently have very bad attacks of cholecystitis and common duct stones. Work with certain inbred strains of mice have revealed a genetically inherited trait that, when fed a high fat diet, they hypersecrete cholesterol into their bile and become stone formers when the cholesterol in the bile precipitates out. It turns out that this gene is very predominant among the Native Americans and indigenous groups of Central America. Not only are they plagued with gall bladder disease, but rates of diabetes and obesity are well over 50 % of the Pima Indian population. It was not always the case.

Previous observations[22] suggest that diabetes was either rare or largely unrecognized among Pimas around the 1900s. At that time, increasing settlement of the area by people of European derivation led to diversion of the Pimas' water supply and disruption of their agriculture. The loss of water resulted in curtailment of subsistence farming and led to fundamental changes in their way of life. In the late 1930s, a review of medical records from the hospitals serving the population identified 21 Pima Indians with diabetes. The authors concluded that the prevalence of diabetes was similar to that in the U.S. population. By the 1950s, many more Pimas were known to have diabetes, and since then, a rising prevalence of obesity suggests that the incidence of diabetes might continue to rise.

Thus with the change in diet from traditional diet high in natural grains and complex carbohydrates to a more Western diet of processed foods with simple carbohydrates and higher fat content, the Native Americans became diabetic and stone formers. Concentrated energy food became more available; the physiologic response to this change was deadly. Why should this be? Plainly speaking, for most species, the response to an increase in available environmental energy would be an increase in population through longer life spans and increased brood size. For the Pima Indians, the physiologic response to this boon of available food energy set up by their genetic code was to kill them off.

Lest you think the Pima are the only ones affected, think again. Throughout the world, emerging societies with changes in the traditional diet to more concentrated caloric Western diets have seen rates of diabetes, kidney failure, and certain forms of cancer sky rocket. Now there are a host of explanations for this phenomenon from increased sedentary activity to inflammatory processes induced by our Western diets, but suffice it to say, it seems our genetic make-up doesn't want us to be too successful. The availability of plentiful foodstuffs could lead to a population explosion. And in the boom and bust cycles of the Hunter-gather world, it is best you not get too big during the boom, lest the whole group perish during the bust. The Pima Indians, living in a marginal desert climate, figured out, genetically speaking, that it would be unwise to expand their numbers when food was abundant, lest everyone should perish when resources revert back to their usual scarce level.

It could be argued that this is simply a byproduct of exposure to an unnatural diet. The native diets didn't expose them to today's type diet so how could they have developed a response to such a diet if the natives have only recently been westernized? In other words the ability to be so efficient at extracting nutrients from the food eaten has only recently been exposed to be a detriment. The increasing incidence of diabetes and gallstones are a consequence of this efficiency not a cause for regulating population numbers.

Yet at times the natives must have been exposed to times of relative plenty and concentrated energy stores. Having the propensity to efficiently process nutrients coupled with a mechanism to prevent overpopulation during times of relative abundance in an environment that is mostly nutrient poor would be an extremely efficient adaptation.

CHAPTER 6

DEPRESSION, SUICIDE AND SQUID CHASERS

Is man the only animal that commits suicide? It is arguable whether a beaching whale or a senescent elephant who wanders off to die is volitionally and consciously ending its life. The spider *Stegodyphus lineatus* allows her young at two weeks of age to feed on her flesh (matriphagy) killing her. That she does not run away makes the act seem suicidal, however, she is not involved actively in killing herself.

Figure 10. The spider Stegodyphus lineatus, allowing her offspring to cannibalize her body. From the website htpp//biology.au.dk/fileadmin/www.biology.au.dk/images/stegodyphus_lineatus_matriphagy.jpg. Photograph courtesy of Trine Bilde.

Man seems to be the only animal that is reproductively viable and consciously and premeditatively ends his own life. Even the misunderstood lemming, does not commit suicide, but is driven by deeply ingrained instincts to migrate when species numbers increase, sometimes to the edge of cliffs where driven by urges to seek out new habitats, they jump or are pushed off into the sea to sink or swim in the struggle to fan out.

In 2005 over 32,000 people in the United States committed suicide. That's slightly more than 1 in 10,000 people who purposefully ended their own life. It's unknown how many people attempt suicide, but is estimated that for every suicide death there are 8 to 25 people that attempt the process.

This is not a new concept. There is evidence of suicide in Egyptian manuscripts that it existed over 4000 years ago. Nor is it culturally specific. Suicide occurs in virtually every culture from hunter-gathers to advanced societies and on every continent. Historically, reaction to suicide has run the gamut from expected to accepted to abhorred.

What is going on? Is it possible that there is a locus on the human genome responsible for suicide? If so, why on Darwin's green Earth would it persist? One cannot fathom an argument in which taking one's own life becomes a survival advantage.

One could argue that complex behavior is not determined genetically. And indeed it does seem such a complicated human act would not be present on a single gene or perhaps group of genes.

Or is it? There are many avenues of evidence that there is a genetic component to suicide. The brain is essentially a tangle, yet a coordinated, highly evolved tangle, of neurons or nerve cells. We have pleasure centers in the brain that have evolved to point us in the right direction: towards rich, high nutrient foods, towards finding a mate and procreating, and perhaps toward taking risks. (i.e. hunting a mastodon, etc). It is easy to see how this would evolve. Those who got active pleasure centers for fatty foods might survive better in a prolonged fasting period. Those with active pleasure centers for risk taking behavior might have more meat on the table or might search out new but unknown avenues of food procurement. That's why when you eat a piece of chocolate perhaps you get a warm fuzzy feeling all over and then proceed to eat the whole box. Fat, sugar and flavinoids in chocolate are a pleasure center trifecta.

We also have pain centers in the brain. Deep in the most primitive portions of the human brain, are areas that are extremely sensitive to noxious stimuli. This area is activated in response to danger and can arouse the "flight or flight" defense.

Most of these neurons in these areas communicate or network with other neurons using a chemical called serotonin. Because we possess this complex tangle of neurons, it is perhaps easy to see that things don't always work seamlessly. We have all heard of the new agents of anti-depressant medications, Prozac™, Zoloft™, etc. Well these drugs increase the concentration of serotonin in the brain and in these pleasure centers which makes us feel better, less depressed.

Depression runs in families. You are 3 times more likely to develop depression if one of your parents had it than if they did not. What could be the survival advantage of developing depression, especially if it may shorten your reproductive lifespan if it persists and leads to suicide?

Depression and empathy likely come hand in hand. Depression probably emerges when we start to care about the people around us. This caring or empathy probably has a survival advantage in that it tends to strengthen the kinship bonds between us, strengthens the clan and improves survival in this way. When bad things happen to a feeling, caring individual, sadness develops and that ability to be sad makes our group more tightly knit. Or it could be a mechanism allowing an individual to pull back from a stressful environment, to process information and live to carry on another day. To carry that notion to its ultimate extreme, suicide, seems inherently unDarwinian. Tendencies toward suicide should not persist in the genetic record, as they should be weeded out over time.

However, there is a lot of evidence that suicidal behavior has a genetic tendency. One study, published in the *Journal of the American Medical Association* in 1985 examined the rate of suicide among the Amish population in southern Pennsylvania. Genealogical and medical records revealed that four families accounted for 73 percent of all suicides, but represented only 16 percent of the total Amish population

Epidemiological studies also back up this idea. For example, two countries that top the world's suicide rate list are Hungary and Finland, with 40 suicides per 100,000 people. Although the countries lie 1,600 kilometers apart, their people share a language group and, presumably, genes. The Finno-Ugric people lived together for thousands of years in the Ural Mountains of what is now Russia, then migrated to Finland and Hungary.

David Bakish, a psychiatrist at the Royal Ottawa Hospital in Canada has studied patients who have an irrepressible thought pattern towards suicide or towards harming themselves and have a family history of suicide. He and his colleagues found that these patients had a mutation in the gene encoding the serotonin 5-HT2A receptor. Serotonin, remember, is the neurotransmitter in the brain, responsible for integrating and relaying information from the pleasure centers in the brain. Errors in signal relay of the serotonin system can have profound implications in mood, particularly depression. Drugs like Prozac™ and Paxil™ keep serotonin around longer which allows the pleasure center's message to become stronger and overall, the mood of the patient improves. Patients with a defect in the gene encoding the 5-HT2A receptor were more than twice as likely to attempt suicide as those with depression but without the mutation. Furthermore, researchers who analyzed the brains of Hungarians who committed suicide showed an overabundance of these serotonin receptors in the brain. If there is not enough serotonin available, the brain will try and compensate by making more receptors in order to catch every last drop of the precious neurotransmitter. Thus genetic errors in either making serotonin or processing it have been associated with to suicidal behavior.

So if there is a genetic component for suicide, could it have evolved in other species besides *Homo sapiens*? Let's examine that whale or dolphin that beaches inexplicably. There are 10 species of whale that commonly beach, all of which are toothed. None of the baleen whales are regularly involved in mass beachings. These are all social marine mammals that appear to be involved with beaching. The tightly knit groups inhabiting deep waters are the most susceptible. Those animals that are solitary seem immune to the phenomenon.

Some have postulated that there is an injury or disease affecting a leader of a pod and that these mass beachings are the result of a tight knit group blindly following or perhaps rescuing or assuaging a sick or confused leader. Some have postulated that SONAR from Navy ships has damaged the brain and rendered these animals unable to echo-locate or process information clearly. Indeed some mass beachings have followed Navy exercises and autopsies have revealed damage to brain tissue and internal organs. Some have postulated that these occur frequently on shallow beaches where the sonar signals are not relayed properly and they are "blinded". But that doesn't explain why the whales are in these shallows to begin with (not their usual locale) nor explain strandings in other locales with deep beaches. Some have postulated a geomagnetic cause.

However, there are numerous instances of mass beachings before the advent of SONAR, so this cannot be the only factor. Furthermore there are beachings now that are not associated with SONAR activity and the whales seem adamant to beach themselves. Well meaning humans have been frustrated on numerous occasions by whales that beach and are successfully returned to the sea with a rising tide or great human effort, only to rebeach two and three times.

SCHOOL OF BLACK-FISH STRANDED ON THE SHORE OF CAPE COD, MASS.

Figure 11 Mass stranding of whales on Cape Cod in 1902. Photo courtesy of National Oceanic & Atmospheric Administration (NOAA)

The whales that are commonly associated with mass strandings (sperm whale, pygmy sperm whale, melon headed whale, beaked whales, pilot whale, Atlantic white sided dolphin) **all** rely on squid as their primary source of food. However, the most common whales involved are the sperm whale and pilot whale. Both are highly social and until 150 years ago had no real predators. They both live off squid whose population is subject to extreme variation. In fact, during periods of La Niña when global ocean temperatures change, populations of squid plummet. When oceans warm again the numbers can explode. Thus the squid-eating whales enjoy a smorgasbord of food punctuated by times of scarcity which can last 3-4 years. These animals live a long time, in large social networks, and are menopausal.[23] Beaching would be a way of keeping numbers in check especially during El Niño periods when squid numbers were high, followed shortly thereafter by La Niña period when squid numbers have plummeted. A study in 2005 by Evans et al. showed a distinct 11 -12 year periodicity in cetacean strandings that correlated well with periodic changes in wind patterns and sea level pressure gradients, which herald changes in resource availability.

These animals are very tightly bonded and likely have close feelings for one another. They look after one another's young while diving in a staggered fashion for food. They have midwives that assist with the deliveries of other mothers. They look after members of their pod who have been injured or stranded by vocalizing and being in close proximity. Their capacity for empathy is great and thus their capacity for depression is likely also great. It would be interesting to check for genetic defects in the serotonin receptor system in pilot, sperm or other whales, especially in the leaders of the those pods that have mass beachings. One could argue that their slow gestation and prolonged calving times tend to keep population numbers in check. However, these animals have long lives, historically little or no predation, tight social structures and a relatively low infant mortality rate all of which could quickly outstrip squid resources especially during prolonged scarcity. Other top predators like lions or humans have relied on aggression and homicide when numbers expand too much. Whales seem ill-equipped for such measures. So, when the winds of climate alteration signal change, a stimulus is interpreted by susceptible cetaceans towards the ultimate act of altruism.

Other theories of mass whale strandings don't add up. Sonar or other acoustic trauma doesn't explain why strandings occurred before such implements were used, nor why some stranded whales have no identifiable cause and seem perfectly healthy. Infectious disease or parasite infestation doesn't adequately explain why only certain species of toothed whales beach, nor does it explain the neither periodicity of occurrences nor the beachings of unparasitized animals.

It may seem ridiculous to ascribe suicide to non-human species, as suicide implies a conscious choice. The human that opts for suicide has seemingly weighed the pros and cons of existing and decided that the latter outstripped the former. However, at its most base, it is a behavior and not always a reasoned one. There seem to be people who have a drive, a fascination, an unquenchable desire to harm themselves. This impulse is then rationalized by humans. In other words, it is the impulse that comes first, the innate behavior, the genetically determined drive for suicide manifests itself and we as humans use our logic and reasoning to explain why we have committed the act. Those with the gene(s) for suicide are perhaps genetically driven to commit suicide, in the same way we are driven to alcohol or risky behavior. They then try and explain why they did it with a note or message. It seems a conscious decision but in fact it may be, for some, a genetically determined proclivity. In the process of developing the complex brain required for social integration, something goes astray. The evolution of whale or human society leads the brain to "overempathasize" in trying to develop a cohesive band. This leads to depression as we know it and ultimately to behavior that is counterintuitive like suicide. In the same way, the whale does not choose to seemingly commit suicide; it is a genetically determined behavior.

If we postulate that suicide has a genetic component then perhaps it is subject to genetic patterns of inheritance, with varying degrees of penetrance in the affected offspring. Those with enough copies of the gene or genes for suicide, it may be enough to contemplate and complete the act. For those with only one copy, mere contemplation might follow. Most are aware of the great minds in the history of the world were subject to fits of depression; Abraham Lincoln, Winston Churchill, Van Gogh, William Faulkner, Ernest Hemingway, Franz Kierkegaard and on and on. The gene for suicide might persist in a population in the same way sickle cell or CF persists; because not only is there a possible advantage to having a proclivity towards intelligence, empathy and depression, but also because there is improved fitness for the individuals where the gene for suicide is found in their peers. The sacrifice of one means more fitness for the rest. The act of suicide need not be a reasoned behavior, only a behavior that results in improved fitness for those that don't carry the full expression of the gene(s) for suicide.

Strength for this argument comes from data to suggest that climate plays a factor in the rate of suicide of a population. The further north one progresses, the higher the rate of suicide. One might postulate that the dearth of sunshine in these climes would lead to a peak of events in wintertime. Paradoxically though most studies have noted a peak in suicides in spring time, in May in the Northern hemisphere and November in the Southern hemisphere. The timing remains mysterious; however, the fact that there exists a peak of events speaks to a genetic mechanism for the behavior. Were the behavior a learned one, one would expect a random distribution of the events throughout the year. It seems that sunlight has some influence on our hormonal processes that leads to a predilection for suicide in susceptible people. Corroborating this notion is the lack of a peak of suicides in equatorial countries where sunlight is relatively constant.

It goes against the grain to suggest that suicide could be a beneficial adaptation, when its expression results in the death of the gene that encodes for the behavior. However, like we have seen with other adaptations, an allele can become widespread, despite the detrimental effect of the full expression of the allele's genetic message if the silent or partial expression of the trait is advantageous.

Chapter 7

HOMOSEXUALITY, HUNTINGTON'S
AND A PARENT'S HELL

Homosexuality, if taken to the extremity of exclusivity, is genetic suicide. Though few animals elicit the behavior exclusively, the practice is common enough that any notion that this is aberrant behavior should be rethought. How do we square this behavior with natural selection and its requirement for increased survival of offspring, when the gene for the behavior should extinguish itself in just a few generations?

According to most researchers, man is not alone in his proclivity for the same sex. Other animals do engage in homosexual behavior as members of bachelor bands, but most are not exclusively homosexual. In recent studies, scholars have identified 450-1500 species that engage in same-sex sexual behavior (Poiani and Dixon). Approximately 8 % of sheep (rams) will choose to engage other rams even when ewes are available. Animals that have lifelong courtship patterns like penguins, have been known to have same sex partners engage in incubating orphan eggs and raising infant penguins.

One could argue that homosexuality is a learned or acquired trait in humans. However there is mounting evidence (could not possibly resist that one) that genetics play a substantial role in determining sexual orientation. Why would the gene for homosexuality persist and remain as prevalent as it is today? From a strict Darwinian interpretation, it would seem that a proclivity towards avoiding heterosexual procreation would be decidedly unlikely to persist.

Biologists have postulated that perhaps there is some advantage to "recessive homosexuality." In other words having some aspects of homosexuality, (but not the full Monty) makes one more attractive to the opposite sex. The recessive trait persists because it is advantageous in its heterozygous form and only detrimental (evolutionarily) in its full blown form. (Not that there's anything wrong with that.)

There are others that look at kin selection as a clue to the persistence of homosexual behavior. It would allow say a gay uncle to help foster the survival of his nieces and nephews with whom he shares 25 % of his own genetic material without having to invest significant energy in raising his own offspring.

It seems an easy explanation to relegate homosexuality to terms of group selection. It provides an outlet for sexual drive that does not result in more mouths to feed. It is good for the group to have males about who do not devolve the group into instability by competing for females. The group or band that fosters this trait will be leaner and more fit to deal with scarce environmental resources. They will be able to outcompete other bands that do not have this trait who are more socially unstable and whose population will be less likely to sustain a prolonged period of dearth. We have seen though the problems with group selection.

Homosexuality could be explained in terms of increasing individual fitness by fostering the drive to procreate but eliminating self-detrimental effects of the deed. So long as the individual is not exclusively homosexual this would hold up. Indeed no other animal except man is exclusively homosexual. And in man there are degrees of homosexuality from the bisexual, to the curious, one-time experimenter, to the situationally homosexual (prisoners, sailors), to the once-married-only-to-find-out-later homosexual.

Imagine that the gene (or genes) for homosexuality has essentially an autosomally recessive inheritance pattern. That is, like cystic fibrosis or sickle cell disease, it takes two copies of the gene to be expressed in the offspring. The heterozygous condition would result in a heterosexual offspring. A promiscuous male carrying the homosexual allele would have a certain percentage of his offspring reproductively silent if at least some of his partners carried the allele as well. This would allow improved survival of the rest of his heterozygous heterosexual offspring in subsequent generations when compared to an equally promiscuous male without the gene. Without the gene the possibility of overcrowding would negatively impact his other offspring's ability to survive in a resource constrained environment. As long as more survive in a population with the homosexual gene in its gene pool, than survive in populations without it, the gene will prosper. Furthermore, a heterozygous heterosexual male may be more attractive towards females if he is in touch with his feminine side. No one is claiming that the homosexual or gene or group of genes is this simple; but there is likely some genetic component to homosexuality at some level, and some form of this inheritance pattern could lead to individual fitness in a fashion described above.

The examples of detrimental traits illustrated in this work are only the tip of the iceberg of examples of genes that are perplexingly malignant, in the concept of natural selection. Almost every organ system has the potential to curtail our lives. Poor eyesight, poor dentition, asthma, food allergy, indifferent maternal instincts, menopause, kidney failure, hepatitis, juvenile arthritis, mental illness, muscular dystrophies, alcoholism, deafness, color blindness, baldness, fair skin, Rh factor hemolysis, hereditary spherocytosis, club foot, and autoimmune disease are just a few potentially deadly traits that are either directly or indirectly the result of genetic mutations that have persisted despite being detrimental to the fitness of the individual who inherited them. However, these traits are not detrimental to the gene(s) itself, in that the frequency of the allele of the gene in question will increase in a time of resource constraint. In other words those with only one copy (or limited expression) of the "maladaptive" gene benefit from the presence of the gene in the gene pool because the full expression of the gene keeps numbers at or below the carrying capacity of the environment. It allows control of gene numbers at both ends of the birth to death spectrum. The adaptive quality of seemingly maladaptive traits is a more plausible explanation than the notion that all of these conditions have not been eliminated because they have been sheltered from the population as a whole and not yet weeded out.

HUNTINGTON'S DISEASE

Huntington's Chorea (Greek for dance) or Disease (HD) is a lethal genetic disease transmitted in an autosomal dominant inheritance pattern. In other words, if your father or mother has the disease, you have a 50% chance of inheriting the disease as well. If both have a single copy of the genetic defect, you have a 75% chance of inheriting the disease. The reason this deadly disease has persisted is that the onset of the symptoms of the disease are not evident until after sexual maturity, usually around 35-50 years of age. Unlike CF or sickle-cell disease where there is a heterozygous advantage, there is no advantage to heterozygous HD, because it only takes one copy of the gene to manifest the deadly disease.

The defect has been traced to our fourth chromosome involving a gene that encodes for a protein called huntingtin, which is a protein found in many neurons in both the brain and spinal cord. The mutant huntingtin contains extra amino acids, notably glutamine, which is thought to activate the apoptosis pathway described earlier. Normal huntingtin has usually twenty or so glutamine repeats. Abnormal huntingtin contains more than 39 but may contain over 100. The extent of the glutamine repeats may be associated with an earlier onset of symptoms. Nonetheless, the abnormally formed protein leads to loss of neuron function in certain vulnerable aspects of the brain and leads to the manifestations of the disease, namely jerky movements, difficulty swallowing, depression, dementia and eventually death.

Because the phenotype is not expressed until after sexual maturity, it would seem that HD is impervious to Darwin's whetstone. The decrement in survival seen with HD would not impair the ability to have offspring so that it would be difficult to rid a population of the defect through natural selection. In fact, some have postulated that those with HD have fewer inhibitions than those without the disease and are more promiscuous. While this is contested by others, investigators have noted that people with HD have 1.14 to 1.34 as many children as their unaffected siblings.[24]

While it is easy to see the benefit of the increase in birth rate would have on the frequency of the gene or allele, it is less obvious what the increment in death would have on the frequency of the gene. However, the average onset of the disease is around 40-45 years old, which would mean that most off-spring of the affected individual would be independent by the time their parent succumbed from the disease. In fact, some would be grandparents or potential grandparents at the time of onset of symptoms. During a drought or in a resource constrained environment, not having a grandparent around (who for all practical purposes is genetically silent anyway) might translate into improved survival of that grandparent's offspring. In the very poor barrios in Barranquitas, Venezuela, the incidence of HD is close to 1 in 10. The prevalence of the disease to this degree is likely not simply attributable to the modest increase in the birth rate. (Remember that only half of children born to an afflicted person will inherit the disease). Though isolation and poverty would be fertile ground to allow genetic drift to continue to foster the disease, some mechanism of increased overall survival (or inclusive fitness) of those with the disease over those without, is in play. I would argue that survival of offspring via propagation of early (but not too early) death under these conditions has enhanced the dissemination of this "pitilessly indifferent" gene. Far from the resource opulence of the West, Darwin's playbook has revealed an ironic closing act; like a mountain climber on a perilously difficult cliff face, the genome of the people of Barranquitas has discovered that timely death can be like a piton hammered into the cliff wall. It allows one's descendants to pull up to the next foothold with enough energy to complete their extraordinarily difficult scale for survival. Those of us with mere uphill saunters by comparison, don't need Huntington's ghastly piton.

CHILDHOOD CANCER

Of all the human maladies capable of befalling us, childhood cancer has to be the most devastating to parent and confusing to the evolutionary biologist. How could a genetic mistake that leads to a cancer that results in death of the child ever get passed on and how could that mistake persist in the genetic record.

There are a host of cancers that afflict children that arise from almost all cell types. The most common disordered cell is the immune cell, specifically the white blood cells. Most are stationed in the lymph system, which if disordered lead to lymphoma. Others circulate throughout the body, which if disordered lead to leukemia. Other cancers involve the brain and bones of children. These are the cells that are rapidly turning over in the growing child. Cells that are rapidly dividing are more prone to mistakes in the transferring of information from parent cell to daughter cell. Childhood cancers arise from mutations in the genes of growing cells. Because these errors occur randomly and unpredictably, currently there is no effective way to prevent them.

Most childhood cancers seem to arise from a genetic defect that regulates growth. A defect occurs that allows these mutant cells to grow unchecked by the usual cellular and subcellular arresting mechanisms. Because often there are redundancies in the DNA command and control centers, often multiple defective genes are required to produce a cancer. Because we get two copies of every gene from both parents, it is unlikely that the defective gene will cause harm in and of itself. This is why childhood cancers are relatively rare. But a viral infection or radiation may be enough to change or inactivate the good copy(ies) in just one cell to allow the cancer to progress.

So there exist in our genome, the seeds of our destruction. These mutant cancer genes are out there (or in there) in the genetic amalgam of our species, producing a very real survival detriment for those that inherit them. Why do they persist? Shouldn't a gene whose expression results in its own death before it can be passed on be eliminated given a certain amount of time?

One could argue that these genes have persisted because they are variably expressed in low numbers. Like sickle cell disease, which requires two copies of the gene to be fully expressed, these genes are piggybacked from generation to generation and only expressed when the planets have aligned properly. And like sickle cell disease, whose heterozygous unsickled phenotypes have some degree of protection from malaria, there may be some unexplained benefit to inheriting a single copy of a "cancer" gene.

Though unusually cruel to contemplate, there is the notion of sacrifice built into our genome. It may seem unfair to ask the smallest among us to participate in that sacrifice. However, during the evolution of our species it was perhaps advantageous if not all individuals survived to adulthood. And it was even more important to accomplish this without reducing our species drive to reproduce. For those that carry these cancer genes, they are survival neutral when incompletely expressed in themselves and survival positive when there is resource scarcity and full expression in others.

CHAPTER 8

RELIGION

For God so loved the world, that he gave his only begotten Son, that whosoever believeth in Him should not perish, but have everlasting life.

John 3:16

One of the most enigmatic quotes from the Bible, John 3:16, comes closest to summing up in a few words, the theme of Christianity. It's why I chose to include it here and why the proselytizing, rainbow-wigged character in the end zone at major football games chose it for his placard. However, sacrifice and the promise of celestial immortality are concepts not unique to Christianity. The idea that some sacrifice their fitness or die young so that others may live is seen in Judaism, Islam, Greek mythology and other religions as well. From the near sacrifice of Abraham's son, Isaac, to the rituals of human sacrifice of the Maya, the notion that some must die so that the rest may go on living is perhaps a process reflecting and reinforcing biologic reality into cultural actualization. In other words, reducing the number of mouths to feed reduces environmental pressure and allows an improvement in survival for those left behind. The actuation of this process, when programmed genetic death is not enough, becomes possible only through a mystical or religious context. The Mayans, at the height of their empire, sacrificed hundreds, even thousands, per day. This coincided with extreme environmental pressure to feed a growing empire. However wholesale slaughter is never received well by the citizenry, unless it is wrapped up as a religious conceptual necessity, i.e. to drive the sun or appease the gods.

The Christian notion of sacrifice is especially compelling in this framework of thought. The idea that God sacrificed his only son so that the rest of Christianity might live is a theme reflected in the notion that early death of an individual, even the most ideal (who is without genetic heirs), is necessary for the survival of those left living. His sacrifice has more ideological benefit obviously than simply decreasing environmental pressure for the rest of us but still retains the symbolic idea that death is necessary for survival of the species.

For the pious, Christ's death is a moral issue; his life and death ultimately represent a path to morality through the promise of immortality. Let's face it, there is often wickedness in surviving. When there is not enough food to go around for everyone, difficult choices are made. Those that eat leave some hungry, and the notion of guilt and sin ensue. Stealing, selfishness, cheating, adultery, and lying all improve my chances of survival over those whom I have sinned against. Furthermore, no matter how unselfish I try to be, there is inevitably someone who will suffer or go without by my consumption of even the most meager of meals. A need to assuage the guilt of the living arises. Christ's death can be seen as assuaging the guilt of survival and the ongoing "wickedness" or exploitiveness associated with survival. Or more palatably, it celebrates the sacrifice of the altruistic.

In myth, children or virgins are often chosen for important sacrifices. Though it might be thought of as an esteemed gift for any deity to receive such a precious commodity, its practice might have been initiated and continued through the benefit the community receives in decreasing population pressure. Food too, is often sacrificed. Thus death and food are comparable. The bread of life for the species becomes death of its individuals. Ritual sacrifice is a theme developed to reflect the necessity of death, even death at a young age. The constant concern of foodstuff scarcity prevalent in pre-industrial times was an overriding cultural concern especially in the desert of the Middle East. However, it is more difficult for us to recognize in our era of relative opulence.

Natural selection of our early ancestors led to an emphasis on fertility and success at reproduction. We are fertile year round. To have survived our early crucible of descent from the tree tops, it was crucial to have evolved an emphasis on replenishment. At some point, with tool use and increased cranial capacity, we perhaps got too good at replenishment. It is perhaps why religion or religiosity is so prevalent today. We have evolved from ancestors who by believing in a higher power were able to keep population levels at a survivable level. Whether this was through the altruistic aspects of ritual communion and sacrifice or the belligerent aspects of religious intolerance is of little consequence; both aspects led to improvement in survival of a population pushing the bounds of what the environment can support.

Religion in today's society seems to be losing its *raison d'être*. Scientific epoxy has replaced the religious oakum that once filled the gaps in our understanding of the universe. Once religion provided a framework for explanation; now it has been relegated, by most, to a moral playbook or social lubricant. The hypocrisy and fallibility of religious leadership further winnows our faith in the institution.

What religion or spirituality can imbue, with a little forward thinking, is an understanding of the processes of life and death and the betterment of ourselves and each other. Morality is important not for the selfish reward I will achieve celestially, later on, but for the reward my children, my heirs, and ultimately my own genes will receive in having a better life for themselves and their heirs. Morality based on religious precept is ultimately selfish, that I will be rewarded or punished for acts good or bad; I shall act morally for my own good later on. Spirituality, for lack of a better word, based on natural selection is ultimately unselfish, in that I act morally for the betterment of my heirs. Placing others needs before my own increases the likelihood of their survival; avoiding socially devolving acts lessens the likelihood of retribution and anarchy, and improves my heirs' survival in a complex society.

Ironically, religion offers the hope of immortality for selfish gain, which is in diametric opposition to the concept of altruism, as hopefully I have satisfactorily explained. I would argue that perhaps it is more important to be recycled into the fabric of our species than to worry about a celestial reward or Stygian torment. We do this by acts of altruism and ultimately by dying. The good earth has a limit of organisms it can sustain. Technology, medicine, sanitation and modern agricultural techniques have pushed that envelope but we are reaching (or have reached) our Malthusian paradigm. Estimates of world population soar to 8 billion by 2025 from 6.6 billion today. If we don't control our population now, the fiends of our darker nature will find an unpleasant solution.

On the other hand, the evolution of malevolent tendencies, so prevalent in our species, begins to make sense. Bellicosity, violent behavior, and cruelty are understandable; these behaviors have been ingrained into our genetic fabric to keep population levels low so that individuals and their genes may survive. Just as those populations which are more altruistic do better in the long run, those populations which exhibit destructive behavior are more likely to persist because they have not outstripped their ability to survive in a resource constrained environment. That we have placed a stigma associated with these behaviors speaks to the angels of our better nature that have developed to keep our large society from falling into chaos and anarchy. But remember codified laws are relatively recent concepts in the grand scheme of human evolution. Nor does this suggest that cruelty is acceptable. However, it lends reason to why it is so prevalent.

The idea that death of some is good for the rest is dangerous ground, I recognize. It is a slippery slope that has led to such nefarious programs of eugenics, racism, genocide and sterilization. Ironically, it is genetic variation that is responsible for the robustness of our species and is the very target of rapacity, genocide and war. Something in our collective past has fostered distrust in that which is foreign. That distrust unfortunately, translated into a survival benefit that is outmoded but still captive in our collective psyche. The more enduring traits, morality and altruism, have allowed our complex society to take hold.

What a reflection of natural selection is culture. That to survive and persist as a species we have to be both bad and good. Good and evil, yin and yang, no wonder these concepts are so prevalent in our group thought (literature, myth, religion). They arise from our primal existence and have both been required for our breeding band to survive. Altruism and cruelty both keep population numbers at or below the carrying capacity of the environment.

When someone dies at a young age from cancer or war or reckless behavior it's often heard that the Lord works in mysterious ways or that we cannot know God's plan. At the risk of biting into the apple, the plan, in my mind, is beginning to unfold. For millions of years, life on this planet has been, from the ergonomic armchair of today's technologically infused society, unimaginably difficult. Humans have been hunters and gatherers in a resource scarce environment and have been throughout our early evolution extremely vulnerable to predation. But we have successfully adapted by socially banding together. By nurturing close bonds, we developed a cultural notion of love, self-sacrifice and rampant altruism. We have developed a breeding strategy that is always receptive and capable of rapid population rise. Furthermore, our developments of intelligence and tool use have drastically decreased predation. However, being too successful or too fruitful puts an individual's genetic legacy at risk, especially in times of widespread drought or calamity. The Indifferent Evolver has preserved the right to keep the tree safe by pruning even healthy branches.

In the end, it perhaps sounds antithetical to Darwin's views to say an individual who shoots for fewer offspring will be better off than the individual with more. But it is not the one that has the most offspring that wins Darwin's race; it is the one that has the most offspring that survive to have their own offspring who wins. Adaptations like altruism, violence, senescence or cancer that delete some genes to save others in the next generation may seem costly; however nature, historically, has bargained from a position of strength. Until very recently, food resources have been scarce and survival has been difficult on the poorly provisioned, storm plagued decks of the H.M.S. Mother Nature. How much would you pay for a flotation device on a sinking ship that could save your children?

Life on earth is not a reproductive sprint to the finish, but a marathon relay without end. Sometimes less is indeed more.

PART III

The Act of Dying

Chapter 1

EXPRESSIONS FOR DEATH

From the number of euphemisms for the term for dying, it is evident that people in this country don't even like mentioning the word death.

Augured in
Bought It
Bought the farm
Bit the dust
Croaked
Passed
In a better place
Crossed over
Paid the boatman
Grew wings
Cashed in his chips
Kicked the bucket
Pushing daisies
Bought the pine condo
Go the way of all flesh
Toes up
Goes Belly up
Met his Maker

Checked out

Sent to Davy Jones Locker

Six feet under

Deceased

Expired

Last gasp

Fell off the twig

Shuffled off the mortal coil

Dead as a doornail

Worm food

Some are obvious metaphors for ending things; some are physical or spiritual descriptions of what happens or what is thought to happen but infused with levity; and some have an obscure meaning that likely was at one time obvious, yet persist having gained a foothold in our collective conscious.

"Kicking the bucket" is an interesting example. "Bucket" is thought to be a bastardization of the French word "buque" (boo-kay) or bar which was employed to hang a carcass prior to butchering it. Thus, if you kick the buque or bar, you are in the last throes of existence.

"Bought the farm" is thought to refer to G.I.'s who purchased combat insurance, should they die, would allow their next of kin to pay off the existing mortgage on the farm. "Poor bastard got shot, but at least he bought the farm back home." Or perhaps it's simply a wry reference to the burial plot that needs to be purchased; the small plot is usually the only "farm" most can afford.

To" expire" literally means to breathe out. Often before dying, the respirations are shallow and slow allowing carbon dioxide (the exhaust fumes of cellular metabolism) to accumulate; the drive to expel carbon dioxide from the body leads one to take a deep breath. If someone is on the very cusp of dying, that last big breath robs the brainstem of what little oxygen is available and the breath becomes the patients last. The last big expiration is his last and he has "expired" in a last gasp effort.

Why a doornail would be the epitome of lifelessness is a little mysterious, however an internet explanation I came across suggests that doornails used to be hammered through the door to add strength then bent backwards on itself to further secure it and keep it from slipping. Iron nails used to be an extremely valuable commodity and were often saved and reused. Thus, bending it backwards rendered the nail unable to be reused, or "dead."

Lest you think health care workers are immune to such euphemistic high jinks, think again. Resident physicians, in the trenches on the front lines in the battle against the Grim Reaper and his minions, are prolific at inventing sobriquets about death. When someone is deathly sick and a group of physicians are discussing his care, you might hear someone request an orthopedic consult. To the uninitiated, getting a consult on a dying patient without a fracture or getting input on patient care from a sawbones seems puzzling and will prompt a quizzical look or remark. To which the physician will reply that he needs a hinge placed surgically in his spine so he can kiss his own ass goodbye. Sardonic physicians have been known to facetiously order a pine box to the bedside, as if the coffin were as necessarily mundane as a box of tissue or bedside slippers.

Patient's that die unexpectedly are said to have "boxed". As in, "What happened to Mrs. Jones, did she transfer to the ICU?"

"No, she boxed last night."

To have died in a code blue with a thousand alarms going off, amidst a great hullabaloo is to have "boxed spectacularly."

A celestial discharge is a nice way of saying it, even though most bodies are discharged to the morgue which is always in the basement of any hospital and never very celestial. A patient that arrived in the emergency room with only fleeting signs of life is termed to have "come in dead and stayed dead." Someone who exsanguinates in the O.R. has "bled out on the table," and the expression "Well, all bleeding eventually stops," is meant to gently taunt a surgical neophyte or colleague who has developed a reputation for ham-handedness. Patients who are not doing well and likely to die soon are said to be "circling the drain." Or "Better get an outpatient pathology consult," means that an autopsy is imminent.

I suppose we invent these terms to allay our qualms about death and dying. In the hospital an untimely death or any death on our watch is very difficult to deal with and infusing the situation with humor perhaps helps a little. Like a callous on the hand of a laborer, cold, crude metaphoric language hardens the emotional repository of our psyche to allow us to deal with death when it is premature. But sometimes we see death as a blessing. When a patient's cancer has spread uncontrollably and further toxic therapy would be futile, a wry consulting oncologist might say, "Give him a fishing pole and six pack therapy." In other words let him live his last days doing what he enjoys using "organic" remedies and not in the sterile hospital getting chemotherapy with his hair falling out. When I've made teaching rounds with my junior colleagues on a patient who has an incurable disease who seems to be suffering or is completely comatose but the family wants everything done, more than once I've heard, unapologetically, "What she needs is a good heart attack."

We don't see people die anymore. It is usually very peaceful. Let me spend a few minutes telling you about a patient who died a good death. This is an unpublished short story I wrote about an unforgettable patient.

Well, there are certain people in medicine you don't forget. He was one of them, a happy-go-lucky hypomanic dude straight from 1967, "Peace, love and Hell no, we won't go". Though he had gone, as a medic. And it affected him. Not so much insanely affected him, only he got a little wanderlust, couldn't sit still, couldn't hold on to a job because he just got restless. Sometimes when he closed his eyes the unspeakable would return and it just plain wore him out.

He used music and art to cope. He could picture a scene and put it down on paper almost like a photograph. And he could play guitar. Post-op day number two, he was sitting up in bed playing Neil Young. Not too strong and a little raspy but there he was.

As he got older he settled down a little, moved back to God's country in the hills of central Ohio far from the madness that accompanies the periphery of that state. He lived from hand to mouth, did odd jobs, had a big garden, built a log cabin, learned the old ways, and studied the Indian lore. He grew a Union infantry man's mustache, big and bushy, and wore an old kepi that such that if you met him in the woods, you'd have thought he was lost from his regiment near Antietam. He had a girlfriend, Alicia, the love of his life, you could tell. She was there for everything, even though it was a long, long way to come. He came in with a bowel obstruction. He hadn't previously had surgery before and that worried me. He wasn't terribly sick, just obstructed. He did not want a colonoscopy, and I respected his concern over something being "shoved up his arse". We waited a day or two, ran a few tests; it looked like a right sided large bowel obstruction, which in the absence of something completely obscure, usually meant cancer. I got to know him very well. We talked for a long time about the Civil War, about where he was from, where he lived, the caves and hills he had explored, the music he had experienced, the herbs and roots he used to affect minor cures.

We operated subsequently and my fears were confirmed. Nine times out of ten I would have brought up a colostomy, or a stoma, a bag for the bowel to empty in at the skin level. The tenet in surgery is to never reconnect bowel that hasn't been cleaned out, for fear that the large number of bacteria in the colon will cause the connection to leak or develop an abscess around where the connection has been made. So you bring up a colostomy and go back later when you can prep the bowel. I didn't want to give my friend a colostomy. We had talked about the possibility beforehand, and he really didn't want any part of that. I left it at that, not really explaining that it would be temporary and not all that bad. So we took out the cancer and reconnected the bowel.

Thankfully, he did fine after that, like I said, playing the guitar up in bed, that drove the nurses crazy and was part of the reason why he did it. I saw him back for his initial check, and the pathology report of the resected specimen showed the cancer had spread to his lymph nodes which meant chemotherapy. So we put a portacath in his vein to allow chemotherapy access to his system and he took it for 6 months. He finished it and I dutifully took the portacath out.

I kinda forgot about him for a while, he would come back every 6 months to get checked up on, CT scans, blood work and the like and everything was clean; then about 2 years from the original operation he came back, obstructed again.

I thought that it would simply be a little scar tissue or band that sometimes causes an obstruction even a long time after an operation, so when his bowels didn't open up, we explored him again. When we opened the last layer before getting inside the belly, a gush of clear fluid, ascites, poured out. On the bowel and caked in the omentum, were the unmistakable gritty, hard implants of cancer, metastatic disease, everywhere, kinking his small bowel, plastering the diaphragm and pelvis. We resected the obstructed bowel and put him back together. Then I took the long slow walk to talk to the family. Alicia was there, or Sunshine as he called her, only. I sat her down and floored her with the news. Absolutely the most gut wrenching, heart-breaking conversation I've ever had. We both never saw this coming, blinded by hopes and dreams. In fact they had made plans to wed. How long does he have? Hard to say, maybe a year. Sometimes the chemo works on these tumors and they have newer agents now…. (Something like that, it's now a blur.)

Well, he did hang on for about a year. The newer chemo agents did slow things a bit. They did get married. Then he came back, obstructed again. I was hoping to help him with a stent or another operation but he developed pneumonia and floated in and out of consciousness. We discussed further treatment but they rightfully declined. Shortly thereafter, an ambulance carried him home and with the help of hospice and good friends and neighbors, they placed him on the covered porch of his log cabin, wrapped in quilts and waited. After two days, he died, a good death, with Alicia right there holding his hand on the front porch with the Ohio sun, filtered through the hemlock, in his eyes.

There is no medical slang term for this kind of death. Amongst all the sadness in the last chapter of his life, his family was comforted by how he died. His was simply a good life preceding a good death.

Chapter 2

DO NOT GO GENTLE

Do not go gentle into that good night,
Old age should burn and rave at close of day;
Rage, rage against the dying of the light.
Though wise men at their end know dark is right,
Because their words had forked no lightning they
Do not go gentle into that good night

From *Do not go gentle into that good night* by Dylan Thomas

I have seen death up close. I have seen it come on cat's paws like Carl Sandburg's fog and I have seen it tsunami-like, unexpected, unwelcome and devastating. I have made a career of bestaying its inevitability. Many have asked but few have answered the conundrum of why we die; like taxes, death is just one of those things that must come to all. Thinking about the reasons we die has led me on a journey of understanding, not only of evolutionary theory, but our concepts of culture and religion.

Americans have an unrealistic notion of death. Modern medicine has prolonged the average life expectancy to the point where we assume it is a right to live to be one hundred. Our attitude towards the inevitability of our own persistence leads us, as a culture, to mistrust physicians, to be litigious, to engage in unhealthy habits, and to vehemently pursue untried, unreasonable, expensive interventions. Recently, the political climate in the United States surrounding end of life issues became almost combustible involving the very private, gut-wrenching decision of a husband to withdraw support from his wife in a persistent vegetative state. Furthermore, the debate surrounding a single payer system of health insurance was derailed by the stigma of "death panels" where left- field concerns from the right of our political spectrum arose that panels of doctors would decide who gets care after a certain age or for certain conditions. Worries were focused on the idea that someone could "pull the plug" on grandma. No one in America seemed alarmed that grandma was "plugged in" in the first place.

This may sound a bit cruel, a bit too close to Dr. Jack Kevorkian for most people's taste. But our attitude towards the sanctity of life and staving off death at all costs leads us to a host of problems. Unnecessary suffering of those without quality of life (or hope for quality of life) is my biggest concern. It hurts to have a tube in your nose or worse yet, through the wall of your abdomen so nutrition is available. IV's, blood draws, and urinary catheters are painful. Respirators are suffocating and uncomfortable. Imagine a forceful gust of air into your lungs when you're not prepared for it. If not physically painful, imagine the indignities of daily life, were one not able to perform the necessary functions of our private lives. Because we are an affluent society we are able to afford, by and large, institutional care for our relatives near the end of life. However, we only see them once the have been cleaned up and settled in. We don't see the surgery for the tracheotomy tube or stomach tube. We don't see the bedsores or soiled bedclothes. We, as a society, are as ignorant to the processes of prolonging life of our elderly in these institutions as Marie Antoinette was to the plight of the peasants of France.

Not only is this process painful, it is expensive. Health care in general is expensive. The currency of natural selection today is not good genes but good health care. And good health care depends on real currency. Rationing health care is a four letter word. But we as a society have to decide what we are going to pay for. Health care rationing goes on everyday, under different guises. In time of war, it is called triage. Insurance companies deny claims. Physicians prescribe generic drugs instead of expensive new therapies. Rational decisions (not rationing) toward end of life care would go a long way in reducing health care expenditures. We have to do better and we can.

In-hospital issues surrounding the end of life are difficult. Mostly decisions are centered around whether a chronically ill or elderly patient, should his or her heart stop beating, have a team of physicians jump start the heart that has failed. On the surface this sounds like a good idea, but if you peel back the covers on the practice, especially for folks with terminal disease, I think most would be shocked at what happens during the procedure and the outcomes of those that have it performed.

Cardiopulmonary resuscitation was originally designed to treat those with a reversible medical condition that has caused the heart to stop: unusual heart rhythms that can be reversed by restarting the electrical circuitry of the heart, heart attack victims who have a clot or blockage that can be quickly reversed, a drowning victim or victim of electrical injury. For those patients with readily reversible cause of the arrest and "a heart too good to fail" CPR does work well.

The process itself requires that the blood in the non beating heart be pushed forward by squeezing the chambers between the breastbone and the backbone. The ribcage is compressed forcefully to squeeze the blood out of the heart. To be effective it has to be forcefully done and often the person performing the compressions must climb onto the bed and place most of his weight over the patient's chest for the process to work. Most ribcages in need of CPR can't take the force required and with the first few compressions, one or more ribs are cracked with a gruesome crunching sound. The heart itself can be contused or bruised by the process. While compressions are going on, the patient is completely unclothed while access to the circulation is achieved, usually in the groin area where the artery and vein of the leg are easily accessible. It's never easy either, without significant blood flowing through the artery or vein it can take 8 to 10 sticks to get access. Also, because the patient is not breathing, usually a team is trying to place a tube in the windpipe to allow controlled delivery of oxygen to the lungs. Often the whole scene is tense and chaotic; there are usually 15-20 people in the room, sometimes more, from pharmacists, to pharmacy technicians, to nurses, to head nurses, to respiratory technicians, to physicians, to residents to medical students and so on. If the heart's rhythm needs to be converted into a more normal rhythm, then pads will be placed on the patient's chest and 200 J of electricity will be administered to the patient, enough to violently contract nearly every muscle in the body, sometimes lifting the patient off the bed. If that doesn't work, it's done again. And again, this time upping the voltage to 360 J.

There have been many studies looking at the success rate of resuscitating patients who have had cardiac arrest both in the hospital and outside the hospital. What do you think the odds are that a patient in the hospital will recover after CPR and leave the hospital on his or her own power? Seventy-five percent? Fifty percent? The answer may shock you more than the defibrillator.

For the patients for which CPR and cardioversion were designed to treat, i.e. those patients with good hearts but a bad rhythm, the success rate (defined as discharge from the hospital) is only about 30-35%. Intuitively, if you are younger your odds fare better. For those who "code" with no heart rhythm (flat-line tracing on the EKG) or who code from other (non-cardiac) causes, like hypoxia, sepsis, blood loss, etc, the chance of making it to discharge is at best 10 % with the odds decreasing depending on your overall health picture. Nursing home residents can expect 2-4% survival after CPR; and though they may survive, the episode can leave the patient with significant neurologic impairment (Cooper S, et al). There are enough published case studies to prove that patients with certain diseases, such as terminal cancer, septicemia, gastrointestinal hemorrhages and acute stroke, who suffer cardiac arrest have minimal chances of survival. In fact, the survival rate is next to zero (Sandroni C, et al).

Is the general public aware of this? In one survey of one hundred elderly people, researchers found that 81% believed that the success of CPR was greater than 50 %. In another survey of nursing home residents, when respondents were informed the actual success rates of CPR, the desire to have it performed on themselves dropped from 44 to 22%. That number dropped to 5 % when the results of CPR on patients with chronic illnesses were factored in.

I don't mean to be a nihilist. I am not advocating forgoing CPR on the elderly; in fact study after study has shown that outcomes for the elderly without significant health problems are nearly as good as those younger than themselves. However, I think we over estimate the power of medicine to save us when we are at the end of life. We are often confronted with decisions regarding the care of ourselves or loved ones who are approaching the end of life; perhaps there are people who would want absolutely everything done or tried, no matter the cost in emotion or dollar. Then again, there are those who wouldn't mind a modicum of dignity, dying with their ribs intact.

Chapter 3

ROUNDING 3RD AND HEADING FOR HOME

Let me tell you another story about someone I took care of. And lest you think it's a story unique to me or my practice, just ask anyone who works in a hospital and takes care of sick patients and they will tell you a remarkably similar story, perhaps with different names or different underlying conditions, but nonetheless the same story, where common sense was trumped by other motives.

Homer Farnsworth[25] had lived a long life. He was 85 years old and had multiple medical problems. He had been at the Bulge in the winter of '44 and had been as cold as man can be without dying. He knew death and he understood the sublime righteousness of simple pleasures, like a room heated with a wood stove in the quiet of winter. He had farmed, built, helped neighbors, and brought in wholly unnecessary copayments of tomatoes and blackberry jam to his favorite physicians and nurses. Once when I appeared before him harried by the various demands placed upon me in my weekly bath of red tape and regulation, he quipped to me matter-of-factly, "Well Doc, at least no one's shootin' at you."

He was referred to me to evaluate a small tumor in his colon that was cancerous but was not significantly impinging on the caliber of his intestinal tract. His kidneys had no longer worked and I had previously created a fistula (a connection between the artery of the arm to a vein of the arm, to allow the vein to dilate enough to allow kidney dialysis three times a week) in his left arm some time ago to be able to safely dialyze him. His wife had died 5 years ago and he lived alone but close to his daughter and her family. He had congestive heart failure with chronically swollen legs and emphysema from the cigarettes handed out like chewing gum so many years before. He took two handfuls of pills every morning, because he was supposed to; checked his blood sugar now and again when he felt like it or felt light headed.

He was, in the surgical parlance, a "train wreck." I suppose it's so named because if you operate on people like this, it's as if you have a long unobstructed view of a train trying to stop on a slick track. To the unsuspecting passenger on board, everything is seemingly okay until the last moment, when it collides with the barrier, and every part of the train is affected by the impact from the coal car to the caboose in a spectacular hair grabbing, car spilling dénouement. You can see it coming a mile a way and if the train goes down that track, there ain't a damn thing you can do about it except watch and hope.

Homer didn't want any part of the operation. I explained to him and his family that his life expectancy was such that he would most likely die with the cancer of some other cause, (heart attack, stroke etc) than of the cancer, and that operating on him would most likely shorten his life rather than extending any meaningful time left on this earth. But his daughter heard the word "cancer" and heard nothing else I had to say. They left my office arguing, Homer adamant on not having the operation, his daughter just as stubbornly repeating, "Daddy you've got cancer, they got to cut it out!"

I was reconsulted about two weeks later when Homer was admitted for something else and I went by to see him when he was alone. He had changed his mind about the operation and now wanted to go through with it. I asked him what had changed his mind and he replied, unconvincingly, that he just couldn't go on knowing he had a tumor in his colon. So we talked some more, I outlined my objections and he said he knew the risks and was willing to give it a go. I finally asked him about his finances, how much disability he received, what happened to the checks when they were delivered, etc. I turns out his checks went to a joint account with his daughter managing most of his finances for him since he had gotten so tied down to those damn kidney machines. So by hook or by crook he had been talked into the operation. I should have simply refused to operate and referred him to someone else, but I was young and cocky and felt if he had a chance to make it through the operation, it should be me at the other end of the scalpel.

Well, to make a long story longer, he lived up to my initial locomotive assessment. It wasn't a terribly difficult operation. He had had only one previous operation 40 years ago to remove his appendix, and we worked quickly and got the specimen out and his colon back together in less than two hours. He was taken off the breathing machine after he had awoken sufficiently and I nearly dislocated my shoulder patting myself on the back. But by day three, the fluid we had given him began to seep into his lungs, his breathing became labored, and pneumonia set in. We put him back on the breathing machine after conferring with his daughter who assured us she wanted everything done. As usually happens, his condition stabilized; after a few days he didn't get any worse but he didn't really get any better. We couldn't get him off the respirator and we couldn't leave the breathing tube in his mouth where it traverses the vocal cords and prevents him from speaking and damages the cords themselves. He also needed nutrition, but because he was too weak to swallow he needed the food to be delivered directly into the stomach. So if he were going to make it out of the hospital he needed a tracheotomy tube to breath and a feeding gastrostomy tube (G-tube) or stomach tube to eat. He also needed special IV line placed in the big veins underneath his collar bone because all of his veins had been used up. His mental functioning was limited to mouthing unintelligible words, which I swear I thought could be translated to "No more" or "Stop, you're hurting me." We were throwing sand on the icy track to try and slow the train on its downhill inexorable course. The family wanted everything done. The first of the month was right around the corner.

We took him back to the O.R where we put him to sleep and surgically cut his neck below the voice box and inserted the short tracheotomy tube. We also punctured his stomach and slid in the permanent G-tube (gastric or stomach tube). While we were there we changed his rectal pouch and reinserted a new urinary catheter because he still made a little bit of urine. He smoldered along, being fed, watered, oxygenated and dialyzed, existing but not really living for another 10 days. Finally, his heart said enough is enough and began to quiver uncontrollably. And before I could get there, and because he was a "full code" (in other words: in the event his heart stops use every means at your disposal to get it restarted) someone had initiated chest compressions in order to get some blood circulating. At the first downward compression, Homer's 7th 8th and 9th ribs cracked like so much kindling and continued to crunch with each thrust. All the adrenalin in the pharmacy wouldn't bring this heart back, nor any attempt at cardioversion, which is the medical euphemism for running 220 Joules of electricity through your system to jumpstart the heart. He was zapped 3 times anyway with enough current to lift him off the bed. As soon as I arrived, we stopped the insanity, pronounced him dead and noted the time. I called the daughter to tell her the news. She was surprisingly (or perhaps not so surprisingly) non-plussed. "I'm sure you did everything you could," was her ungracious reply. To which I said through gritted teeth, "Yes, we did," but could have easily have been translated as something decidedly less diplomatic.

Not every case is this dramatic, nor are financial incentives the only reason caregivers choose to try and do everything to keep someone alive. Sometimes, people simply think everything must be done just because the technology is there to do it. Sometimes it is the fault of the physician or surgeon who has an incentive to recommend an operation in general, as a result of our procedure-based reimbursement system. Physicians in Great Britain and in our own Veterans Affairs hospitals can perhaps be more perspicacious in selecting patients for operations because they are salaried and not paid based on the number of procedures performed.

The only thing I reasonably plead for is common sense. Understand we are meant to die. No amount of health care spending or technologic breakthrough is going to change this fact. Our expenditure of gross domestic product on health care is approximately 17 % in 2007 and expected to be near 20 % in the near term. Great Britain spends half of what we do per capita and their life expectancy is one year longer!! If we reduced our health care spending to what the average country does, we could expect to save 1 trillion dollars a year! Most of what we spend is in the last few months of life, and most of that spending does not contribute to a higher quality of life. We feel like the humane or Christian thing to do is prolong life for those where the technology exists to be able to do so. I would argue it is inhumane to prolong artificially the life of those which have little or no intrinsic quality. It is tough to decide when that occurs, especially for someone else, so it is imperative to let those around you know your wishes in writing or otherwise. Way, way too often well meaning caregivers prolong life of those who wouldn't want it prolonged.

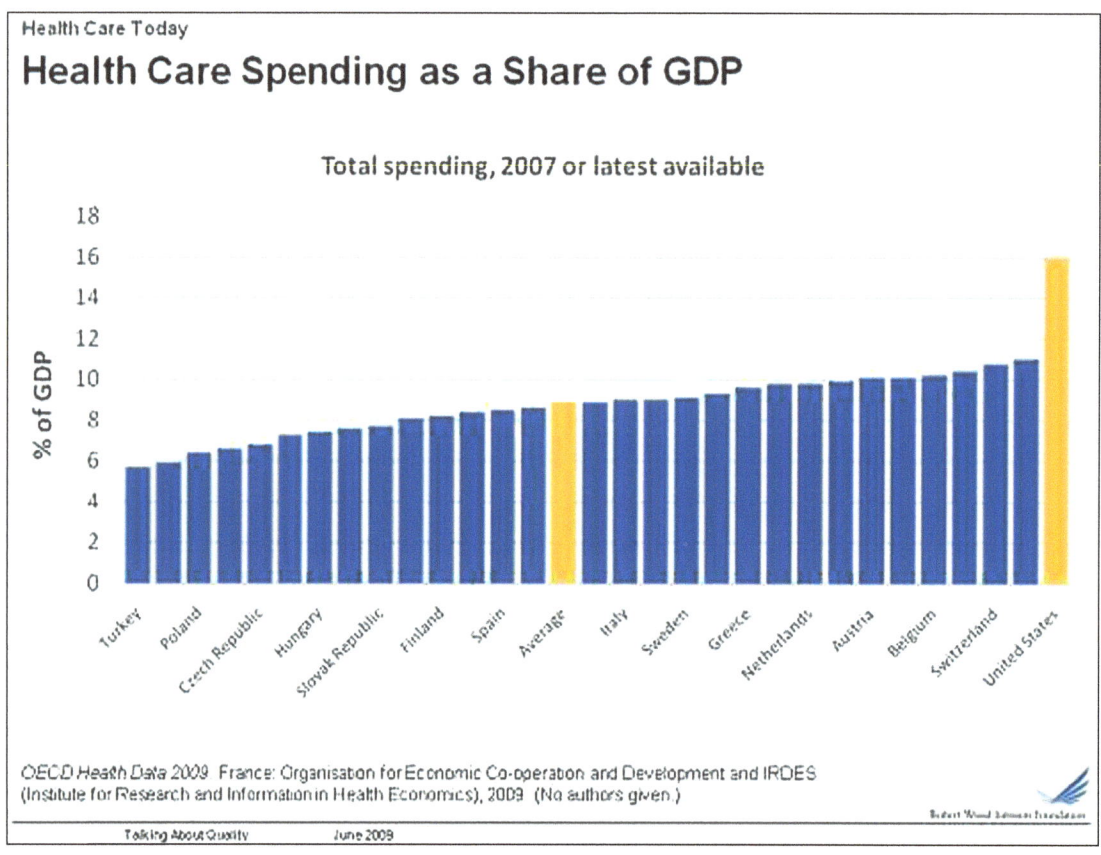

Figure 12. One of these countries is not like the others!!! From the website Robert Wood Johnson Foundation at www.rwjf.org/pr/product.jsp?id=45110. Reproduced with permission of the Robert Wood Johnson Foundation, Princeton, N.J., Copyright 2009.

The men and women of "The Greatest Generation" sacrificed overseas for the rest of us. Subsequent generation sacrificed in Korea, Vietnam and the Middle East. Our grandchildren deserve our sacrifice as well; by controlling costs at the end of meaningful life, we don't suck the marrow from the inheritance meant for our children and grandchildren. I don't mean just monetary inheritance either. We are at the edge of what our planet can sustain in carbon emission and the consequences of human population explosion.

There is no political force available to mandate such a mechanism of sacrifice. It must come from educated, pragmatic, selfless individuals who realize that, inexorably, life on earth is not meant to be forever. And when it is truly our time, we can die with quiet dignity on our hemlock shaded porches or we can run up a big bill for someone else to pay, dying in the sanitized wards of some impersonal hospital. I know what I'm going to do. Do you?

Don't misunderstand me, I am all for using technology to save and extend lives. I am not suggesting a modest proposal for reigning in health care costs. The work I perform has been an uninterrupted crusade against cancer, microbial infection, kidney failure and untimely death. Let us neither throw away the miracles that modern medicine has affected nor give up on those without means but suffer from difficult maladies. We do need to scrutinize, through rigorous, unbiased trials, what we pay for and what we get in return. As physicians, we should not be tempted by the wolves of industry, purveying our white coats as sheep's raiments, dispossessing a flock we swore to protect. We should be circumspect of newer chemotherapy agents that cost thousands per dose, and only extend our chronology on the order of weeks over traditional (and thus less expensive) therapies. We should never accept the word of manufacturers of bright, shiny new drugs as gospel. It's one thing to emulate Don Quixote, struggling against the windmills of inevitability, on a barnyard nag; it's quite another level of foolishness to do so in a brand new BMW.

We also have to be careful for what we wish for. Most everyone, given their druthers, would like to live longer. Research into curbing senescence sounds good on the surface, but could have dire consequences if brought to fruition for the multitudes. The impact of the human race on the globe right now is dire; pushing the envelope of population growth by postponing natural death could be catastrophic for the entire planet. Though our intelligence and tool use have afforded us some respite against the indifferences of nature, we have yet to evolve 20/20 circumspection for evaluating the consequences of our selfish predilections.

Why do we die? Well, because we do. Natural selection has discovered that dying is good for us, as individual genetic torch bearers, in that it allows our offspring to more successfully have their own offspring. If natural selection had put its trust in the long-lived individual for efficient gene propagation, then resources would be diverted from the most vigorous to the least. It's why pine trees make lighter fluid in their sap. It's why annuals fade and elderly Eskimos used to wander off on dark snowy nights.

Thus full circle have we come. Live life every day, for it is a gift, a gift from those who have passed before. Evolution (or whatever is responsible for the pleuripotent, almost magical properties of DNA and chemical bonds and thus evolution, if you prefer) has seen to it that our own death is a gift to those whom we love and leave on earth. But do not go gently into that good night; though realize it is where we are supposed to go. I suppose that is why Mr. Thomas labeled the night "good". Like Ted Williams, the Splendid Splinter, hit a home run in your last at bat, play hard that last game and hope it never ends; but know that peace comes from working hard, playing hard and knowing when it's time to bow out gracefully, without tripping on your G-tube.[26]

APPENDIX

An experiment:

To determine whether or not sexual selection can lead to an increase in mortality of a certain sex, the following experiment might be performed.

Fruit flies (*Drosophila melangaster*) exhibit size dimorphism with the females being slightly larger than the males and easily distinguished.

Three enclosures, of equal numbers of fruit flies, two of which have a death trap are constructed. The death trap is composed of a suction or vacuum with a mesh covering the suction apparatus whose pore size is just small enough that most of the males cannot be sucked out into the suction trap. The suction would be randomly turned on for a few seconds every few hours and should not injure the flies to any degree if they are not sucked up.

The first enclosure has a death trap and food is given *ad lib*.

The second enclosure has the death trap and food is given very sparingly, just enough to barely keep 50% of the flies alive, provided every other day or so. Food should be enough to keep the flies from decreasing body size significantly.

A third enclosure has the same feeding regimen as the second, but there is no death trap.

Flies in the first trap would not be expected to shrink. There should be selection pressures towards increasing the size of the flies.

In the second enclosure, again, intuitively one would expect that selection pressures would favor larger males who would not get sucked up through the holes in the mesh. However, I would hazard a guess that selection pressures would favor females who begin to select for smaller males who would be more prone to be aspirated into the death trap. This arrangement would allow the female and more importantly her offspring more food to survive with.

The third enclosure acts to ensure that starvation alone doesn't decrease body size and thus acts as a further control.

If the male flies in the second enclosure shrink in size faster than the ones in the third or first, then we can reasonably assert that this is due to selection pressure favoring females who mate with smaller males.

Just as a peahen chooses a male who with his large impediment is more likely to die, the fruit fly female also chooses a smaller mate who in this arrangement is more likely to die. In so doing, she improves the fitness of their offspring. Her male offspring may be more likely to die as well, but so long as death is not guaranteed by the death trap (because he could avoid it), but is more likely assured by caloric deprivation, pressure to promote the well being of the offspring at the expense of the parent(s) should be seen.

Bibliography

- Ajdacic-Gross V, Bopp M, Ring M, Gutzwiller F, Rossler W. Seasonality in suicide–a review and search of new concepts for explaining the heterogeneous phenomena. Soc Sci Med. 2010 Aug;71(4):657-66. Epub 2010 Jun 4. Review. PubMed PMID: 20573433.
- Aubert, G and Lansdorp PM. Telomeres and Aging. Physiol. Rev. 88:557-579, 2008.
- Axelrod R, 1984, *The Evolution of Cooperation*, New York: Basic Books
- Bagemihl, B. *Biological Exuberance: animal homosexuality and natural diversity.* Macmillan Science 1999.
- Braude S,Tang-Martinez Z, Taylor GT. Stress, testosterone, and the immunoredistribution hypothesis. Behavioral Ecology (1999) 10 (3): 345-350.
- Clutton-Brock TH, Isvaran K. Sex differences in ageing in natural populations of vertebrates. Proc Biol Sci. 2007 Dec 22;274(1629):3097-104
- Cooper S, Janghorbani M, Cooper G. A decade of in-hospital resuscitation: outcomes and prediction of survival? Resuscitation 2006; 68: 231–7.
- Coyne, J.A. *Why Evolution is True*. Penguin (2010)
- Cronin, Helena, *The Ant and the Peacock: Altruism and Sexual Selection from Darwin to Today.* New York, Cambridge University Press, 1991.
- Darwin, C. *On the Origin of Species by Means of Natural Selection*, London: John Murray,1859.
- Darwin, C. *The Descent of Man, and Selection in Relation to Sex,* London:1871
- Dawkins, Richard, *The Extended Phenotype*, Oxford University Press, 1982
- Dawkins, Richard, *The Selfish Gene*, Oxford University Press, 1976
- Dawkins, Richard, *River Out of Eden: A Darwinian View of Life*, Basic Books, 1995

- Du L, Bakish D, Lapierre YD, Ravindran AV, Hrdina PD. Association of polymorphism of serotonin 2A receptor gene with suicidal ideation in major depressive disorder. Am J Med Genet. 2000 Feb 7;96(1):56-60.
- Egeland JA and Sussex JN. Suicide and Family Loading for Affective Disorders. JAMA 1985;254:915-918.
- Evans K, Thresher R, Warneke RM, Bradshaw CJA, Pook M, Thiele D, and Hindell MA. Periodic variability in cetacean strandings: links to large-scale climate events. Biol Lett. 2005 June 22; 1(2): 147–150.
- Flescher, A.M. and Worthen, D.L. *The Altruistic Species: Scientific, Philosophical, and Religious Perspectives of Human Benevolence.* Templeton Foundation Press, West Conshohocken, PA, 2007.
- Fletcher, J. A. and Zwick, M., 2004, 'Strong Altruism Can Evolve in Randomly Formed Groups', *Journal of Theoretical Biology*, 228: 303-13
- Fletcher, J. A. and Doebeli, M., 2006, 'How Altruism Evolves: Assortment and Synergy', *Journal of Evolutionary Biology*, 19: 1389-1393
- Gould, S. J. *Ever Since Darwin.* 1977. New York: W. W. Norton.
- Gould, S. J. *The Panda's Thumb.* 1980. New York: W. W. Norton.
- Gould, S. J.. *Dinosaur in a Haystack.* 1995 New York: Harmony Books.
- Guynup, S. A Suicide Gene, Is there a genetic cause for suicide? Genome News Network, Online May 12, 2000.
- Hamilton, W. D., 1964, The Genetical Evolution of Social Behaviour I and II, *Journal of Theoretical Biology*, 7: 1-16, 17-32
- Hamilton, W. D., 1970, Selfish and Spiteful Behaviour in an Evolutionary Model, *Nature*, 228: 1218-1220
- Hamilton, W. D., 1972, Altruism and Related Phenomena, Mainly in the Social Insects, *Annual Review of Ecology and Systematics*, 3: 193-232.
- Hamilton, W. D., 1975, 'Innate Social Aptitudes in Man: an Approach from Evolutionary Genetics', in *Biosocial Anthropology*, R. Fox (ed.), New York: Wiley.
- Hamilton, W. D., 1996, *Narrow Roads of Gene Land*, New York: W. H. Freeman.
- Hughes A.L. and Hughes M.K. Paternal Investment and Sexual Size Dimorphism in North American Passerines, *Oikos*, Vol. 46, No. 2 (Apr., 1986), pp. 171-175
- Israel S, Lerer E, Shalev I, Uzefovsky F, Riebold M, et al. (2009) The Oxytocin Receptor (*OXTR*) Contributes to Prosocial Fund Allocations in the Dictator Game and the Social Value Orientations Task. PLoS ONE 4(5): e5535. doi:10.1371/journal.pone.0005535
- Jamison, K.R. *NIGHT FALLS FAST Understanding Suicide.* New York:Alfred A. Knopf. 1999.

- Le Boeuf BJ, Reiter J, 1988. Lifetime reproductive success in northern elephant seals. In: *Reproductive Success: Studies of Individual Variation in Contrasting Breeding Systems* (Clutton-Brock TH, ed). Chicago: University of Chicago Press; 344-362.

- Marlowe, F.W. Hunter-Gatherers and Human Evolution. Evolutionary Anthropology 14:54 –67 (2005)

- Maynard Smith, J., 1998, The Origin of Altruism, *Nature*, 393: 639-640

- Mitteldorf, J "Whence comes death? How natural selection has fashioned aging and death; why "natural" isn't the last word in healthy foods and lifestyles; and some observations on objective science pursued by subjective humans". Humanist. Jan –Feb 2002. Online

- Owens, I.P.F. (1995). Hormonal basis of sexual dimorphism in birds: implications for new theories for new theories of sexual selection. *TREE,* 10 (1): 44- 47.

- Pavkov ME, Hanson RL, Knowler WC, Bennet PH, Krakoff J, and Nelson RG. Changing Patterns of Type 2 Diabetes Incidence Among Pima Indians Diabetes Care. July 2007 vol. 30 no. 7 1758-1763.

- Petrie, M. (1994). Improved growth and survival of offspring of peacocks with more elaborate tails. *Nature,* 371: 598-599.

- Poiani A and Dixson AF. *Animal Homosexuality: A Biosocial Perspective.* Cambridge University Press, 2010.

- Reed WL, Clark ME, Parker PG, Raouf SA, Arguedas N, Monk DS, Snajdr E, Nolan V Jr, Ketterson ED. Physiological effects on demography: a long-term experimental study of testosterone's effects on fitness. Am Nat. 2006 May; 167(5):667-83. Epub 2006 Apr 5.

- Reynolds, J.E. *Marine mammal research: conservation beyond crisis.* 2005. Johns Hopkins University Press.

- Roughgarden*, J. The Genial Gene: Deconstructing Darwinian Selfishness.* University of California Press, 2009

- Roy, A. Genetic and biologic risk factors for suicide in depressive disorders. Psychiatric Quarterly Volume 64, Number 4, 345-358, DOI: 10.1007/BF01064927

- Sandroni C, Nolan J, Cavallaro F, Antonelli M. In-hospital cardiac arrest: incidence, prognosis and possible measures to improve survival. Intensive Care Med 2007; 33:237–45.

- Schneider,J.M., Salomon, M. and Lubin, Y.Limited adaptive life-history plasticity in a semelparous spider, *Stegodyphus lineatus* (Eresidae) *Evolutionary Ecology Research*, 2003, **5**: 731–738.

- Trivers, R. L. 1971 The Evolution of Reciprocal Altruism, Quarterly Review of Biology 46: 35-57

- Trivers, R. L. 1972 Parental Investment and Sexual Selection in B. Campbell (ed*.), Sexual Selection and the Descent of Man*, 1871-1971 (pp. 136□179) Chicago, Il: Aldine

- Trivers, R. L. 1974 Parent/Offspring Conflict, American Zoologist, 14 249-264
- Weatherhead P.J., Robertson R.J. Offspring quality and the polygyny threshold: 'The sexy son hypothesis'. Am Nat. **113** (2): 201–8. Feb 1979; doi:10.1086/283379
- Wilkinson, G. S., 1984, Reciprocal Food Sharing in the Vampire Bat, *Nature*, 308: 181-184
- Williams, G. C., 1966, *Adaptation and Natural Selection*, Princeton: Princeton University Press
- Wilson E. O., 1975, *Sociobiology: the New Synthesis*, Cambridge MA: Harvard University Press
- Wrangham R.W. *Chimpanzee Cultures* 1994, Chicago Academy of Sciences
- Zahavi, A. and Zahavi, A. (1997) *The handicap principle: a missing piece of Darwin's puzzle.* Oxford University Press. Oxford.
- Zak PJ, Stanton AA, Ahmadi S (2007) Oxytocin increases generosity in humans. PLoS ONE 2: e1128

Endnotes

1   *Turritopsis nutricula* is an unusual hydrozoan which can revert back to earlier stages of development and back to the adult stage and in theory could be immortal; however, it has never been observed to have such longevity. There are species of hydra which are thought not to age; however this is disputed by others in the field.

2   You can interpret the phrase "magic gene number" in a number of ways. It might mean the number of an individual gene if you are a gene centrist. It might mean genome number or the entirety of genes that make up an individual. Magic gene number might reflect the gene or group of genes that control reproduction or brood size and thus dictates a species' population size. Generically, the gene in the phrase "magic gene number" refers to a hypothetical gene in an individual's genome, but ultimately it should refer to the gene or genes that regulates lifespan of the organism.

3   These are classic problems that have arisen that test Darwin's theory. These questions have been asked and answered by scientists more capable than I. Perhaps it is not that the theory of kin selection or reciprocal altruism has a hard time explaining these contradictions, but that the answers are somewhat complicated. Perhaps there is a simpler way of explaining these phenomena that complements those complicated theorems.

4   Curiously, most sexual dimorphism in birds is driven by testosterone. The peacock however, if castrated, will still develop his resplendent plumage. If the female's ovaries are removed she will also develop the male phenotype. It may be that the ratio of testosterone to estrogen is important for the tail feathers to develop, such that the relative dearth of estrogen is the driving force of tail maturation. Whatever the mechanism- testosterone in most dimorphic birds or an estrogen-like compound in peafowl- some genetic tendency towards a gaudy tail is being selected for by the peahen.

5   Reed WL, Clark ME, Parker PG, Raouf SA, Arguedas N, Monk DS, Snajdr E, Nolan V Jr, Ketterson ED. Physiological effects on demography: a long-term experimental study of testosterone's effects on fitness. Am Nat. 2006 May; 167(5):667-83. Epub 2006 Apr 5

6   Though it is hard to demonstrate an actual difference in mortality of peacocks versus hens in the wild, one could argue that the larger tail has an actual survival benefit in displaying against predation, making the peacock appear larger and more intimidating. However it has been shown in passerines that the sexually dimorphic traits do correlate with an increase the males' mortality. (Daniel E. L. Promislow, Robert Montgomerie and Thomas E. Martin. (Mortality Costs of Sexual Dimorphism in Birds. *Proceedings: Biological Sciences*. Vol. 250, No. 1328 (Nov. 23, 1992), pp. 143-150.) Furthermore it is conclusively the case with dimorphic mammals. Intuitively it is hard to imagine how the tail could be construed as a survival advantage, but I have to admit that the data is not conclusive either way. If it were an advantage, one might argue that the female should develop the tail as well.

7   Trivers, R. L. (1972) Parental investment and sexual selection. In B. Campbell (Ed.) *Sexual selection and the descent of man*, 1871-1971 (pp 136–179). Chicago, Aldine

8   Hoist by his own petard refers to being blown up by one's own bomb. I suppose this is a suitable metaphor for the peacock who might think of himself or his tail as being "the bomb"

9   Hughes AL and Hughes MK. Paternal Investment and Sexual Size Dimorphism in North American Passerines. Oikos Vol. 46, No. 2 (Apr., 1986), pp. 171-175

10 I don't mean to assign a negative connotation to the male peafowl's mode of parenting. "Deadbeat" simply describes an absent paternal strategy and should not be misunderstood to be a pejorative opinion of the peacock.

11 Loffredo CA and Borgia G. Male Courtship Vocalizations as Cues for Mate Choice in the Satin Bowerbird (Ptilonorhynchus violaceus). *The Auk*, Vol. 103, No. 1 (Jan., 1986), pp. 189-195.
    or
    Siefferman, L., Hill, G. E. and Dobson, F. S. 2005. Ornamental plumage coloration and condition are dependent on age in eastern bluebirds Sialia sialis. J. Avian Biol. 36: 428-/435.

12 Braude S,Tang-Martinez Z, Taylor GT. Stress, testosterone, and the immunoredistribution hypothesis. Behavioral Ecology (1999) 10 (3): 345-350.

13 Le Boeuf BJ, Reiter J, 1988. Lifetime reproductive success in northern elephant seals. In: Reproductive Success: Studies of Individual Variation in Contrasting Breeding Systems (Clutton-Brock TH, ed). Chicago: University of Chicago Press; 344-362.

14 Clutton-Brock TH, Isvaran K. Sex differences in ageing in natural populations of vertebrates. Proc Biol Sci. 2007 Dec 22;274(1629):3097-104

15 Were you worried I was going to use a different alliterative word?

16 Bacterial colonies floating in environments with limited substrate is a reasonable metaphor for more complex animals in their ecosystems, be it bee hive, ant colony, wolf pack or human tribe. Expand too rapidly, and the whole colony or family is at risk for perishing. One might argue that I am carrying this metaphor too far; a group of animals can contract with starvation or predation until the carrying capacity of the environment for the group is reached. There seems to be no dramatic tipping population point of a wolf pack which if reached, all will perish like the floating bacteria. But I think most would agree that there is some number of animals that if suddenly appeared in an already overtaxed ecosystem, that would jeopardize the whole number of animals in existence in that ecosystem. Adaptations like senescence, programmed death, altruism, or genetic disease have evolved to ensure that that number is never reached. Furthermore, the whole colony need not be at risk at one dramatic denouement like a mat sinking; the colony (composed of individuals) that continually over expands by pushing the envelope of environmental sustainability is in jeopardy of being out competed by other reproductive units (families or colonies) that stay under the limit of what the environment can maintain by regulating deaths as well as births.

17 From the website U.S. Army Center of Military History, http://www.history.army.mil/html/moh/iraq.html

18 The tendency of warblers to raise the chicks of brood parasite cuckoos, even to the point where the warbler should realize that the monstrously large toddler in the nest is clearly not of the same species, seems puzzling. Scientists have cogently argued that the instinct to assign kinship to those close by is a rational, if exploitable, Darwinian instinct. It seems odd though that some warblers can recognize a poorly imitated cuckoo egg and abandon it, but not recognize the behemoth fledgling cuckoo. It might be more understandable if this process eases environmental constraint and enhances previous offspring survival, if the warbler "allows" the cuckoo to murder his present offspring. The warbler does not add further to the crowded warbler population by raising a cuckoo that will presumably not compete for the same resources that her previous offspring are vying for. Cuckoos are specialized hairy caterpillar feeders, a resource that most others birds are loathe to exploit. By raising caterpillar killers, this may have a positive environmental impact on the warblers habitat or food resource, that otherwise would be negatively impacted if the caterpillar population got out of control. Obviously, allowing the brood parasitism is not a premeditated decision

Figure. Reed Warbler feeding a Common Cuckoo chick in a nest. Photo: Per H. Olsen, courtesy of Wikipedia.

on the part of the warbler, but any selective pressure to recognize the fledgling cuckoo might be outdone by the benefits of raising a few caterpillar killers.

19 The timing of optimal offspring independence is obviously species dependent. For salmon and other fish this happens at the moment of fertilization. For other species this may be a difficult time to pinpoint. Individuals in a species use natural selection to decide the optimum number of offspring they should have. Having too many offspring above the optimal number will be detrimental to both current and earlier offspring; it might result in species numbers falling below the carrying capacity of the environment and risk selection in favor of less fecund individuals. We should expect this overpopulating tendency to be counteracted with adaptations that reduce population numbers; while reducing fecundity is part of that equation, it is not the only part.

20 For an interesting article that supports this notion at least in bacteria see http://www.nature.com/news/2011/110317/full/news.2011.166.html

21 Marlowe, F.W. Hunter-Gatherers and Human Evolution. Evolutionary Anthropology 14:54 –67 (2005)

22 Pavkov ME, Hanson RL, Knowler WC, Bennet PH, Krakoff J, and Nelson RG. Changing Patterns of Type 2 Diabetes Incidence Among Pima Indians Diabetes Care. July 2007 vol. 30 no. 7 1758-1763.

23 Humans, sperm whales and pilot whales are the only animals in the wild that both have menopause and end their own lives. The reason for menopause is not some adaptive reason like helping with child rearing or foraging, but adaptive in that it keeps population in check in a species that is successful in avoiding predation and other causes of early death.

24 Tufts University (2007, September 26). Biologists Link Huntington's Disease To Health Benefits In Young. *ScienceDaily*. Retrieved June 28, 2011, from http://www.sciencedaily.com /releases/2007/09/070925130029.htm

25 Not a real patient's name

26 If you placed end to end all the G-tubes I have inserted, they would easily span the length of a football field. Some folks need them to eat and go on living; I have no issue with this life saving therapy and would happily place them in any patient who needed it. More often we are asked to place them so people can die. Multiple studies have shown that 1 in 5 patients over 60 die within 30 days of getting a G-tube and only 50 % live more than a year. Often these are severely debilitated patients with no quality of life. Placing the feeding tube allows the caregivers and referring physicians to sleep comfortably at night assuming that they have done all that could be done. If you asked the referring physician, "Would you want this tube yourself were you in the same position?" The answer would be no in a landslide. Death is all too often seen as a defeat; sometimes it is a blessing. Always is it inevitable. (See Angus F. and Burakoff R. The Percutaneous Endoscopic Gastrostomy Tube: Medical and Ethical Issues in Placement. American Journal Of Gastroenterology, Vol. 98, No. 2, 2003, for an excellent review of this topic.)